PHYSICS BY EXPER

AN ADVANCED-LEVEL COURSE

TEACHER'S GUIDE

J. R. L. Hartley B.Sc. (Hons)

and

D. L. Misell B.Sc., Ph.D., F.Inst.P., C.Phys., Dip.R.M.S.

Epsom College, Surrey

Stanley Thornes (Publishers) Ltd

First published in 1987 by:
Stanley Thornes (Publishers) Ltd
Old Station Drive
Leckhampton
CHELTENHAM GL53 0DN
England

British Library Cataloguing in Publication Data
Hartley, J.
Physics by experiment-an advanced-level course.
Teacher's guide
1. Physics—Experiments
I. Title II. Misell, D.
530'.0724 QC33

ISBN 0-85950-192-2

Typeset by Tech-Set, Gateshead, Tyne & Wear.
Printed and bound in Great Britain at The Bath Press, Avon.

CONTENTS

Preface v

Acknowledgements vi

1 An introduction to practical physics 1

5 Introductory experiments 3

A Measurement of the e.m.f. and internal resistance of a cell 3

C Measurement of the specific heat capacity of a metal block using electrical heating 3

6 Mechanics and mechanical properties of matter 4

B Acceleration due to gravity 4
C Collisions 4
D Motion in a circle 4
E Moment of inertia 4
F Rotational kinetic energy 5
G Strength of materials 5
H Elasticity 5
I Viscous forces 5

7 Oscillations and waves 7

A Simple harmonic motion 7
B Damped free vibrations 7
C Damped forced vibrations 7
D Stationary waves in free air 7
E Stationary waves in an air column 8
F Stationary waves on a wire (I) 8
G Stationary waves on a wire (II) 8

8 Geometrical and physical optics 9

Special vernier scales 9

A Refraction at a plane boundary 9
B Converging and diverging lenses 9
C Converging and diverging mirrors 9
D Deviation by a prism 9
E Thin film interference 9
F Diffraction grating spectra 10

9 Thermal properties of matter 11

A Calorimetry 11
B Thermal conductivity (I) 11
C Thermal conductivity (II) 11
D Saturated vapour pressure 11
E Latent heat of vaporisation 11

10 Current electricity 12

A Diode characteristics 12
B Meter connections 12
C Energy transfer 13
General points for potentiometer and metre bridge experiments 13
D Potentiometer (I) 14
E Potentiometer (II) 14
F Potentiometer (III) 14
G Potentiometer (IV) 14
H Metre bridge (I) 14
I Metre bridge (II) 14
J Metre bridge (III) 15

11 Electric and magnetic fields 16

A Hall effect 16
B Ballistic galvanometer 16
C Alternating magnetic fields 16
D Electromagnetic induction 16
E Specific charge of an electron (I) 16
F Specific charge of an electron (II) 17

12 Capacitance 18

A Direct measurement of capacitance 18
B Charging and discharging a capacitor (I) 18
C Charging and discharging a capacitor (II) 18

13 Alternating currents 19

A Capacitive reactance 19
B Inductive reactance 19
C Series resonance 19
D Parallel resonance 19
E Lissajous' figures 19

14 Electrons, atoms and nuclei 20

A The photoelectric effect 20
B Bohr theory 20
C Radioactivity 20
D Radioactive decay 21

Introduction to electronics 22

15 Analogue electronics 23

16 Digital electronics 25

17 Problem experiments 28

Measurement

A Measurement of the radius of curvature of a concave surface 28
B Flow of liquid through a burette 28

Mechanical properties of matter

C Bending of a loaded metre rule 29
D Oscillations of a hacksaw blade 29
E Extension of a spring 29
F Variation of the flow of water through a capillary tube with temperature 30

Oscillations and waves

G Oscillations of a bifilar pendulum 30
H Oscillations of a pendulum against a knife edge 30
I Oscillations of a spring system 30
J Oscillations of a loaded test tube 31
K Damped oscillations of a half-metre rule 31
L Vibrations of a vertical wire under tension 32

Geometrical and physical optics

M Measurement of the refractive index of a liquid 32
N Refraction by a cylindrical lens 32
O Use of displaced images to determine the refractive index of a block 33
P Measurement of the focal length of a converging lens using conjugate images 33
Q Measurement of the focal length of an inaccessible lens 33
R Measurment of the absorption of light in glass using an LDR 34
S Measurement of the wavelength of light using a diffraction grating 34
T Measurement of the optical rotation of polarised light 34

Thermal properties of matter

U Measurements on the cooling of borax solutions 35
V Measurement of the thermal conductivity of glass 36
W Measurements on a vapour in an enclosed space 36

Current electricity

X Measurements on the electrical characteristics of a component 37
Y Measurement of the e.m.f. and internal resistance of a source 37
Z Measurement of the resistance of a galvanometer using a metre bridge 38
AA Measurements on the characteristics of a lamp using a metre bridge 38

Capacitance

AB Discharge of a capacitor through a resistor 39

Semiconductor devices

AC Measurements on the characteristics of an LED 39
AD Measurements on an LED and an LDR 39

Appendix 2 Least squares fitting of experimental data 41

PREFACE

In this guide to 'Physics by Experiment' we have provided detailed information on the construction of any special pieces of equipment required. This particularly applies to the sections on Analogue and Digital Electronics newly introduced into the core of the A-level Physics syllabus. Wherever relevant, we have given advice on components and alternatives, if the specified equipment is unavailable.

Occasionally we have provided additional theory and analysis, which you may wish to give to the more mathematically minded students. We felt that these additions would have been inappropriate in the student's text.

The problem experiments taken from A-level practical examinations have been provided with the 'Instructions to supervisors' as published by the Examination Boards. While many teachers will prefer to use experiments taken from their own Examination Board, we do believe that many of the practicals provide a challenge to students, particularly if they are required to provide a full theoretical analysis of the data.

J.R.L.H.
D.L.M.
1987 Epsom College, Surrey

ACKNOWLEDGEMENTS

While writing this book we have received advice and encouragement from Geoff Camplin, Patricia Hayes, Ian Host, Mark Purdon and David Wright. We have valued the help and support from our wives and families. We are grateful to Janet Misell for writing the BBC BASIC computer program for the least squares fitting of experimental data.

We are grateful to the following Examining Boards for permission to reproduce A-level practical problems from their examination papers:

The Associated Examining Board
Joint Matriculation Board
Oxford and Cambridge Schools Examination Board
University of Cambridge Local Examinations Syndicate
University of London School Examinations Board
University of Oxford Delegacy of Local Examinations.

1 AN INTRODUCTION TO PRACTICAL PHYSICS

In the student's text an oblique reference has been made to experiments and apparatus that the student will not encounter outside a school laboratory. Ideally we should have liked to have omitted many of these experiments, in favour of those experiments that require the use of modern techniques and extensive analysis on the part of the student.

These experiments include:

a) apparatus that takes a long time to set up or takes a long time to reach a steady state. Examples include the measurement of a small e.m.f. using a potentiometer and the measurement of thermal conductivity using Searle's bar or Lees' disc.

b) obsolete apparatus. Examples include the potentiometer and metre bridge, and even moving-coil meters. All these could be replaced by digital multimeters or other suitable electronic devices.

c) experiments involving a large number of measurements of doubtful precision. Examples include collisions of trolleys.

Clearly, many of these can be presented as demonstrations, but on the advice of the assessors, we have included many of these because they are an integral part of many practical courses and the principles at least are required by most A-level syllabuses. Notable casualties which have been excluded are:

6 Mechanics and mechanical properties of matter
Verification of $F = ma$
Flow of liquid through a tube
Surface tension experiments

7 Oscillations and waves
Stationary waves on a string (Melde's apparatus)

8 Geometrical and physical optics
The magnification and resolution of a telescope
Young's two slit experiment

9 Thermal properties of matter
Measurement of specific heat capacity by the method of mixtures
Measurement of the latent heat of fusion of ice
Continuous flow calorimetry (Callendar and Barnes)
The constant volume thermometer
Measurement of the ratio of the principal molar heat capacities of a gas (Clement and Désormes)

10 Current electricity
Use of the potentiometer to measure high voltages
Use of the potentiometer to calibrate a voltmeter

11 Electric and magnetic fields
Measurement of the charge on an electron (Millikan's oil drop experiment)
Verification of the laws of electromagnetic induction

14 Electrons, atoms and nuclei
The inverse square law for γ-radiation
The characteristics of a Geiger–Müller counter
Excitation and ionisation energies (Franck–Hertz experiment)

Against this we set some new experiments, which we believe not only provide an excellent way of investigating physical phenomena, but also provide the student with the opportunity of doing some worthwhile analysis:

6A Coplanar forces: Investigation of the equilibrium of a body under the action of three coplanar forces

6E Moment of inertia: Investigation of the rolling of cylinders down an inclined plane

6I Viscous forces: Measurements on falling spheres in a liquid

7A Simple harmonic motion: Investigation of the extension and vibrations of a loaded spring

7B Damped free vibrations: Investigation of the free oscillations of a moving coil galvanometer

7C Damped forced vibrations: Investigation of the forced oscillations of a moving coil galvanometer

10B Meter connections: Measurment of resistance with an ammeter and a voltmeter, and investigation of the errors that meters introduce

10J Metre bridge (III): Investigation of the variation of resistance of a semiconductor with temperature

13C Series resonance: Investigation of current and charge resonance in an L–R–C series combination

The titles may be familiar but we hope that we have achieved a new approach in all these experiments.

The major addition includes analogue and digital electronics (sections 15 and 16), which cover the requirements for all the new A-level physics syllabuses.

1.1 THE PURPOSE OF EXPERIMENTAL PHYSICS

The recommendations given in the student's text on presenting their experimental work are not intended to be prescriptive. However, we have found that this is often an area neglected by students and very good results are not presented in their best form.

1.2 HOW TO USE THIS BOOK

We have found that a few introductory lessons on experimental uncertainty and graph plotting are useful before the student starts any experimental work. The material in sections 2 and 3 is intended to provide a detailed course on these items. It is not expected that all the material should be covered, but rather that the student should refer to the appropriate paragraphs during the complete course. Section 4 provides six exercises in data and graphical analysis based on experimental data obtained by us. Thus the final results are not expected to be in exact agreement with theoretical or accepted values.

In order to delay the start of A-level experiments until a significant amount of theory has been covered, and also to provide an introduction to a more analytical approach to practical work, we have included three introductory experiments in section 5. The apparatus required for these experiments is usually available in class sets and it was our intention that the whole class should do these experiments at the same time. This also provides the opportunity for class discussion after the practical session.

With the exception of the analogue and digital electronics in sections 15 and 16, it is expected that the other experiments are used in rotation within a group of students. Each experiment includes a section 'Principles involved' to cover the possibility that the student has not covered the theory in class.

For the analogue and digital electronics, it is expected that the students will be split into two groups. Then each group will work singly or in pairs through the complete series of experiments. The groups would then switch over after a specific number of sessions. The time allocated depends on the A-level syllabus taken, but it is likely to be a continuous period of 2–3 weeks. As presented, you should not need to spend a significant amount of time introducing analogue and digital electronics.

As far as the main experiments are concerned we have included, where relevant, the following information in this book:

a) details of any particular equipment required that is thought to be non-standard.

b) details of equipment that requires preparation or construction.

c) alternative apparatus.

d) theoretical analysis that is considered inappropriate in the student's text but which you may find useful to have available for curious students.

The student has been provided with a set of Hints and Further Guidance at the back of the student's book. We cannot prevent students using these as if they were a set of answers, but we hope that they will be encouraged to use these aids only if they wish to check a doubtful step or value. Often, students can learn by making errors in the analysis of their results, and often they can rectify these errors by having available the aids we have given on graph plotting, units, approximate values of physical quantities and hints on answers to questions within each experiment.

There are three other items provided for the student:

Appendix 1 A practical guide to setting up a CRO.

Appendix 2 A BBC Basic computer program for the least squares fitting of experimental data by a straight line. This program also gives values for the uncertainties in the slope and intercept.

Appendix 3 Tables of selected physical quantities appropriate to the experiments in the book. We felt that this would encourage students to look up values for quantities that they have measured and make an objective comparison. It is not intended to be a cheap replacement for Kaye and Laby or any other data book. But often these comprehensive data books are quite daunting to use and the less comprehensive data books omit some of the quantities we have included. The tables are arranged in the same sequence as the experiments in this book.

Finally, we have provided a set of problem experiments in section 17 based on past A-level practical examinations. In this book the original instructions provided by the Examination Boards have been included. Clearly, you may wish to use only practicals from your own Board, but we do believe that these experiments can be usefully given throughout the course. The preparation is, in many cases, quite simple. Some in fact provide the opportunity for first class experimentation and analysis; as such you may consider giving these to the most able students as projects, requiring them to produce the theoretical analysis not required in the examination.

5 INTRODUCTORY EXPERIMENTS

5A MEASUREMENT OF THE e.m.f. AND INTERNAL RESISTANCE OF A CELL

An artificial resistance is added to the cell for two reasons: firstly, the slope of the graph of V against I cannot be measured sufficiently accurately for $r = 0.1$ to $0.2\ \Omega$ (a typical value for a dry cell). Secondly, the current drawn from the cell for low values of the load resistance R is more than 100 mA and the cell life will be considerably reduced.

A resistance of $10\ \Omega$ fitted in series with the cell gives a reasonable slope for measurement and limits the maximum current through R to about 150 mA. Even this limits cell life. The cell life can be increased by using a $100\ \Omega$ resistor for r, limiting I to about 15 mA. A larger value for R of about $500\ \Omega$ will then be required with a 0–10 mA ammeter.

The cell can be mounted on a wooden block with a 0.5 W $10\ \Omega$ RS resistor soldered to one terminal. 3 4 mm terminals are fitted to the block, so that the cell can be used with (1 and 3) or without (2 and 3) its series resistor (Fig. 1).

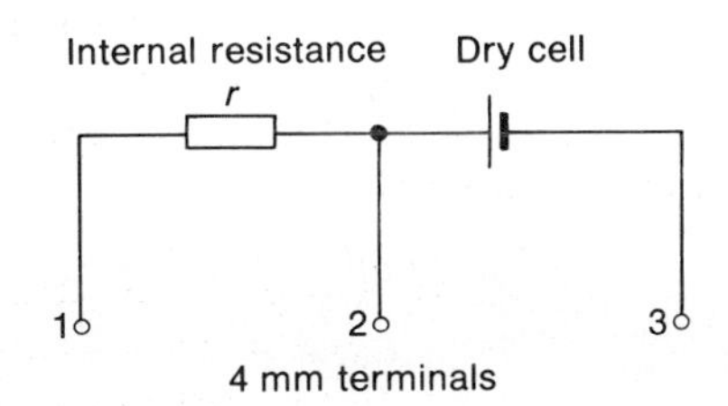

Fig. 1.

5C MEASUREMENT OF THE SPECIFIC HEAT CAPACITY OF A METAL BLOCK USING ELECTRICAL HEATING

If an aluminium block (Nuffield Year 2) is not available, then a brass ($c = 370\ \mathrm{J\,kg^{-1}\,K^{-1}}$) or iron ($c = 480\ \mathrm{J\,kg^{-1}\,K^{-1}}$) block may be used. The heating time should be reduced as appropriate compared with the aluminium block ($c = 880\ \mathrm{J\,kg^{-1}\,K^{-1}}$).

Some 50 W, 12 V immersion heaters are too long for the hole in the block. Although the heating element is at the bottom of the immersion heater, there is still a significant energy loss from the exposed top of the heater.

d and **p** Newton's law of cooling is valid under conditions of forced convection or in a steady draught. Empirically the rate of energy loss in still air is proportional to the excess temperature to the power of 5/4. However, using this cooling behaviour leads to a more complicated analysis for the cooling correction and its calculation in terms of the ratio of areas A_1 and A_2 under the $\theta - t$ graph would not be valid.

In the derivation in (p) $mc(\mathrm{d}\theta/\mathrm{d}t) = k\theta^{5/4}$ or $\mathrm{d}\theta = K\theta^{5/4}\,\mathrm{d}t$.

The integrations now give $\delta\theta_1 = \int_0^{t_1} K\theta^{5/4}\mathrm{d}t$ and

$$\delta\theta_2 = \int_{t_1}^{(t_1+t_2)} K\theta^{5/4}\,\mathrm{d}t.$$

These integrals are not simply related to A_1 and A_2.

These integrals are in fact the areas that would be obtained if a graph of $\theta^{5/4}$ against t were drawn. Students who are more mathematically minded may be asked to try this form of analysis and see what effect it has on the cooling correction.

It is often stated that the differences between the θ and $\theta^{5/4}$ cooling laws are negligible for small temperature excesses θ. The difference in cooling rates is quite significant: $\theta^{5/4}/\theta$ is 1.78 at 10 K, 2.11 at 20 K and 2.51 at 40 K excess temperature.

6 MECHANICS AND MECHANICAL PROPERTIES OF MATTER

6B ACCELERATION DUE TO GRAVITY

The electromagnet can be made by winding 28 SWG or 30 SWG enamelled copper wire on a wooden former; the respective resistances per unit length are 0.155 and 0.221 $\Omega\ m^{-1}$. The final coil should have a resistance of about 40 Ω, which limits the current to about 50 mA using the centisecond timer.

Some timers may not be able to cope with the back e.m.f. developed at the break of the electromagnetic circuit. A diode should then be included in the electromagnetic connections to the timer so as to prevent a reverse current.

Timers with the terminals as shown in Fig. 6.11b in the student's book cannot handle changes in the 'stop terminal' status if they occur before changes in the 'start terminal' status.

A wall mounted version of the apparatus is preferred because of the problems encountered in lining up the ball-bearing with the shutter. But it is essential that the distance of fall should be variable between 1 and 2 m, so that the students can complete the graphical analysis.

6C COLLISIONS

This is a tedious experiment, with the measurements taking most of a practical double period. There are some important principles of mechanics involved but it is suggested that the student should perform the complete experiment for only one type of collision. Observations on the other two types of collision would be made, without any attempt to obtain quantitative data.

In selecting the trolleys it is preferable to have two of the same mass, otherwise certain parts of the experiment, (j), (q) and (w), cannot be completed satisfactorily.

6D MOTION IN A CIRCLE

It is very difficult to achieve accurate results using the procedure described. The alternative radial railway track apparatus is not standard equipment in most schools.

Even if your normal practice is for students to do experiments on their own, this experiment really requires two students, one to concentrate on rotating the rubber bung in a circle with a constant speed and the other to count and time the revolutions. A nylon monofilament with a breaking strength of 30 N is not strictly necessary. But it does have two advantages: it is virtually impossible to break no matter how fast the bung is whirled, and the friction between the tube and the nylon is quite low. Cotton thread does tend to fray during repeated use.

Note that the alternative use of a spring balance to measure the tension in the monofilament is not recommended. Students find it very difficult to obtain a constant reading and, in addition, a correction must be made for the actual weight of the spring balance.

6E MOMENT OF INERTIA

The preparation time required to prepare two cylinders of identical mass is considerable, but it is considered that the effort is well rewarded by students seeing the importance of mass distribution in determining the dynamics of rotating bodies.

Preparation of the cylinders

To obtain two cylinders of identical mass and external radius only is quite simple, provided that different lengths can be used for the cylinders. However, students do find difficulty in accepting that the differences in the lengths are irrelevant. To produce cylinders of identical mass M and external dimensions R and l a steel (or iron) pipe is used for the hollow tube and a car body filler is used for the solid cylinder.

a) Hollow cylinder: the pipe should have an external diameter of 45–50 mm and a wall thickness of 3–5 mm. If its wall is much thinner than this it may be impossible to make its mass as large as that of the solid cylinder. If the wall is too thick then metal can be machined from the inside using a lathe.

b) Solid cylinder: the recommended filler is david's P.38 ISOPON (resin + filler) whose density, when made up, is just right for achieving equality of mass with the pipe. It is made up in a Perspex or glass tube with an internal diameter greater than or equal to that of the pipe after any initial rust/paint has been removed. The filler can be easily cut and machined down to the correct external diameter on a lathe.

c) Calculation of the dimensions of the hollow cylinder: it is useful before machining the inside of the pipe to work out how much material needs to be removed from inside the pipe. If the internal diameter of the pipe is r, then for equality of mass:

$$\pi R^2 \rho_s = \pi (R^2 - r^2) \rho_h$$

for cylinders of equal length. The density of ISOPON, ρ_s, is 1690 kg m^{-3} and the density of steel, ρ_h, is 7860 kg m^{-3}. From this equation it can be shown that $(r^2/R^2) = (\rho_h - \rho_s)/\rho_h$ or $r/R = 0.886$. If r/R is greater than 0.886 it will be impossible to achieve equality of mass; you need a pipe with a larger wall thickness. The required wall thickness $(R - r)$ is $0.114R$. For example if $R = 25$ mm and the wall thickness of your pipe is 5 mm, the wall thickness must be reduced to

$$0.114 \times 25 = 2.85 \text{ mm}$$

d) Machining the hollow cylinder: the nearer you can get to the required internal diameter without machining, the better, because steel pipe must be machined very slowly. The material must be removed evenly from the inside, otherwise the pipe will not rotate smoothly. When you are near to the required wall thickness/internal diameter, remove the pipe from the lathe and check its mass against that of the solid cylinder. It is now a matter of trial and error to achieve masses that are the same to within ±5 g. Once a single set has been produced, subsequent sets can easily be made.

6F ROTATIONAL KINETIC ENERGY

A wall mounted flywheel is preferred. Note that if its moment of inertia is significantly greater than 0.02 kg m^2 and the frictional torque greater than 0.002 N m, the falling mass must be at least 100 g. In this case it is suggested that students use 100 g steps for m up to 500 g rather than 250 g.

So that the experiment is more than just an exercise in measuring the moment of inertia of the flywheel, students have been asked to calculate the frictional torque of the flywheel bearings by two methods. Some students may find this to be too mathematical. It is therefore suggested that the additional timing in (g) to determine the time for the flywheel to stop can be omitted, together with the calculations in (n) and (o).

6G STRENGTH OF MATERIALS

You may prefer to use a horizontal version of the apparatus as shown in Fig. 2. It does have the advantage that greater lengths of material can be used with correspondingly larger extensions. It is not quite as easy to use for extensions in excess of a metre, which occur for natural rubber.

Hoffmann clips are used as an alternative to the proper clamps for securing the sample; these clamps (for example Griffin and George XBV-650-H) are very expensive.

With the exception of the copper (whose diameter is just about right), try to find samples of the other materials with as small a cross-sectional area as possible. Many of the samples listed require a load in excess of 10 N to obtain a complete force–extension curve to breakage. The only material difficult to acquire is the natural rubber strip with a small enough thickness (less than 0.5 mm). Some suppliers of biological materials offer sheets of natural rubber which can (with difficulty) be cut to size with a razor.

6H ELASTICITY

Steel wire or piano wire of the required diameter is usually available. However, if you use iron wire or copper wire instead you should inform your students of the relevant tensile strengths: for iron $\sigma_T = 300$ MN m^{-2} and for copper $\sigma_T = 140$ MN m^{-2}. Otherwise they may use loads that exceed the yield strength of the wire.

6I VISCOUS FORCES

A Perspex tube is preferred to a glass one. Students could crack a glass tube when removing the ball-bearings from the bottom of the tube with a strong magnet.

To complete this experiment satisfactorily it is necessary to have a wide range of steel ball-bearing sizes, about 6 ranging from 1.5 to 8 mm in diameter. Exact metric sizes for some of these diameters are available from standard suppliers. However, a complete range is available in imperial sizes from K. R. Whiston Ltd, New Mills, Stockport SK12 4PT (minimum order is 100 of each size).

Recommended sizes are 1/16″ (1.6 mm), 3/32″ (2.4 mm), 1/8″ (3.2 mm), 5/32″ (4.0 mm), 3/16″ (4.8 mm), 1/4″ (6.4 mm) and 5/16″ (7.9 mm).

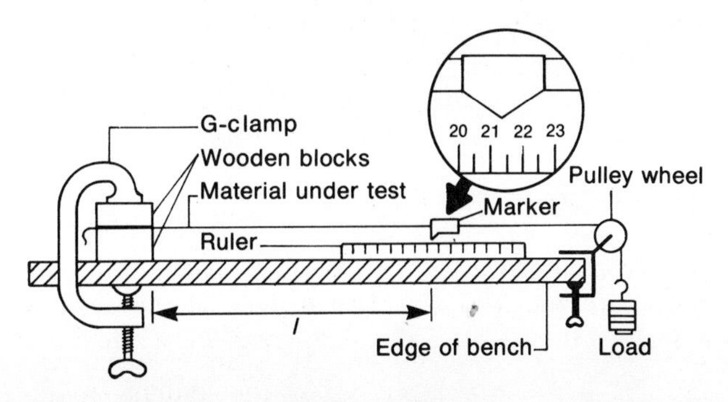

Fig. 2. Alternative experimental arrangement for measuring the behaviour of materials under stress.

The use of a 10 mm diameter ball-bearing is not recommended, since calculations show that there is a significant amount of turbulence caused in its fall through glycerol. The results for this sphere diameter will therefore deviate significantly from the theoretical result.

b The derivation of the curves for a and v is quite standard using the equation of motion: $m\mathrm{d}v/\mathrm{d}t = mg - kv$ with the upthrust term omitted. The solutions for a and v are $a = g\exp(-kt/m)$ and $v = (mg/k)[1 - \exp(-kt/m)]$.

However, of rather more use experimentally is the equation relating the distance x fallen and the speed v of the sphere. This enables a calculation of the distance x_{min} in which the sphere achieves its terminal velocity v_{t} within a certain limit, for example, $v = 0.999\,v_{\mathrm{t}}$ or within 0.1% of v_{t}.

Derivation of the relationship between x and v_t

(neglecting upthrust)

The equation of motion above is rewritten as:

$$mv\frac{\mathrm{d}v}{\mathrm{d}x} = mg - kv$$

since $a = \mathrm{d}v/\mathrm{d}t = (\mathrm{d}v/\mathrm{d}x).(\mathrm{d}x/\mathrm{d}t) = (\mathrm{d}v/\mathrm{d}x).v$.

Rearranging this equation enables the integrations with respect to v and x to be made:

$$\frac{v\,\mathrm{d}v}{\left(g - \frac{k}{m}v\right)} = \mathrm{d}x$$

giving:

$$x = -\frac{m}{k}\left[v + \frac{mg}{k}\log_{\mathrm{e}}\left(g - \frac{k}{m}v\right)\right] + \text{constant}$$

The constant of integration is obtained by using $v = 0$ when $x = 0$: constant $= (gm^2/k^2)\log_{\mathrm{e}} g$.

The final equation for x is:

$$x = -\frac{m}{k}\left[v + \frac{mg}{k}\log_{\mathrm{e}}\left(1 - \frac{kv}{mg}\right)\right]$$

Suppose you now wish to find the value of x for which $v = 0.999v_{\mathrm{t}}$. The value of v_{t} is approximately mg/k. Therefore substitute in the equation for x, $v_{\mathrm{t}} = 0.999mg/k$ to give:

$$x = -\frac{m^2g}{k^2}[0.999 + \log_{\mathrm{e}}(1 - 0.999)]$$

Considering the data for the largest sphere ($r = 4$ mm) and taking a low value for the viscosity of glycerol of $1\ \mathrm{N\,s\,m^{-2}}$: $k = 6\pi\eta r = 0.0754\ \mathrm{N\,s\,m^{-1}}$, $m = 4\pi r^3\rho_{\mathrm{s}}/3 = 2.09 \times 10^{-3}$ kg; thus $v_{\mathrm{t}} = mg/k = 0.272\ \mathrm{ms^{-1}}$ and $m^2g/k^2 = 7.53 \times 10^{-3}$ m. The value of x is then 0.044 m. Thus, by 0.05 m the largest sphere has reached its terminal velocity.

In view of this analysis there did not seem to be much point in asking students to measure the times of fall of the spheres between two sets of markers, so as to establish that the terminal velocity had been reached. This diminishes the accuracy with which v_{t} can be determined (assuming the same length of tube is used) and detracts from the main purpose of the experiment, which is to investigate how v_{t} varies with r.

h Students have, on occasions, found that the sphere is actually moving slower with the marker further down the tube. Of the two possibilities, the variation in the density of glycerol with depth, and the variation of the viscosity of the glycerol as a result of temperature gradients, the latter reason is the most plausible. A temperature variation of as little as 1 K can cause a 7% change in the viscosity of glycerol. The temperature of the column of liquid can vary by 0.5 K and this easily accounts for the unusual behaviour. Shaking the column of liquid is not recommended as the bubbles caused take some time to rise to the top of the liquid column.

n and **s** There is in fact a far more elegant way to determine η and at the same time eliminate any effects arising from the finite size of the sphere. A graph of v_{t}/r^2 against r should be a horizontal line with a constant value of $(2/9\eta)(\rho_{\mathrm{s}} - \rho_{\mathrm{l}})g$ (see equation [6.18] in student text).

However, because of deviations from the theoretical behaviour, particularly at large r values, this graph is a curve with a negative gradient. It is nearly horizontal for small values of r and the graph can therefore easily be extrapolated to $r = 0$. Thus an ideal value of v_{t}/r^2 can be calculated, and the value of η free from edge effects and any turbulence can be determined.

7 OSCILLATIONS AND WAVES

7A SIMPLE HARMONIC MOTION

The spring constant should be between 15 and 20 N m^{-1} for the loads used in these experiments. The length of the spring is not important unless you would like your students to investigate the coupled vertical and horizontal oscillations observed for a particular value of the load. To observe the transition from vertical to horizontal oscillations, the length of the spring must be between 0.25 and 0.35 m. For shorter springs the coupled oscillations occur for very small loads. Suitable springs include: Griffin and George XBV-480-C with an unstretched length of 0.33 m and a force constant of 16.7 N m^{-1}.

7B DAMPED FREE VIBRATIONS

A suitable light beam galvanometer is the WPA 'EDSPOT' with an undamped period of about 2 s.

The resistance values given are based on a galvanometer resistance of $R_g = 14\ \Omega$. The critical resistance for critical damping is usually marked on the back plate of the galvanometer; it is approximately $9 \times R_g$. For over-damping the resistance R_d should be approximately R_g.

The resistor R_p is made up from two linear carbon track potentiometers with values 0–100 kΩ and 0–1 MΩ. A fixed 10 kΩ carbon resistor (0.5 W) is included in series with the 0–100 kΩ potentiometer, so as to prevent a large current accidently going through the galvanometer (Fig. 3).

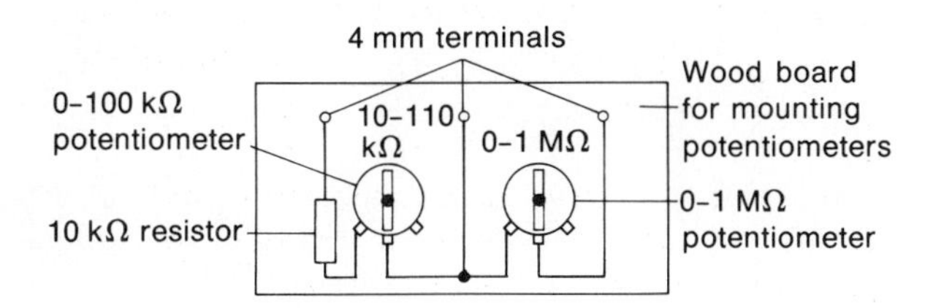

Fig. 3.

The capacitor should be chosen so as to give a full scale deflection for a potential difference of 2 V. The charge sensitivity of the galvanometer is usually marked on the back plate of the galvanometer; it is about 80 mm μC^{-1}. Thus for a full scale deflection, Q = scale length/charge sensitivity. So $C = Q/V$ can be determined. For the standard EDSPOT $Q = 90/80 \approx 1\ \mu C$ and $C = 0.5\ \mu F$. The nearest standard value for C is 0.47 μF.

Note also that the time constant for the capacitor–resistor circuit must be significantly less than the period of oscillation of the galvanometer suspension. For normal settings CR_p is about 20 ms, which is much less than the period. Thus the capacitor discharging through the galvanometer can be considered to provide an instantaneous impulse.

7C DAMPED FORCED VIBRATIONS

The same EDSPOT should be used as in experiment 7B.

e The resonance should occur at almost exactly the value of f_0 determined from the period of free oscillation of the galvanometer. Frequently, the calibration of the scale on the signal generator can be in error by as much as 10%. Thus if students find that the frequency as determined by the signal generator is significantly different from f_0 all subsequent frequency readings should be corrected by the appropriate factor assuming a constant percentage error; for example, if f_0 is measured as 0.5 Hz, but the signal generator reads 0.45 Hz when resonance occurs, then all frequency readings should be multiplied by 1.11. You may not think it necessary to trouble students with this correction as it does not alter the relative shapes of the curves. The correction does ensure that the resonance curves are consistent with the expected value for the resonance peak f_0.

7D STATIONARY WAVES IN FREE AIR

A crystal (piezoelectric) microphone is recommended because these have a high sensitivity (about 50 dB) giving an output of about 50 mV on a CRO. Higher quality unidirectional microphones cannot be used without an amplifier.

The CRO should have a Y-plate sensitivity of 20 mV div^{-1}.

Note that the microphone actually detects pressure nodes and antinodes, but this does not affect the measurements of the wavelength of the stationary waves. But it may worry students that there is a maximum displayed on the CRO when the microphone is near to the reflecting barrier.

7E STATIONARY WAVES IN AN AIR COLUMN

The version with the moveable tube in preferred. The other version can be quite tedious to use, requiring small adjustments to the water level using a second beaker of water.

7F STATIONARY WAVES ON A WIRE (I)

Most of the figures given apply to a sonometer wire with a diameter d of 0.46 mm (26 SWG). The maximum tension in the wire should be altered if a thinner wire is used: P_{max} is proportional to d^2. Table A3.4 in the student text lists P_{max} for a range of d values.

The length l_0 for the wire corresponding to a frequency of 256 Hz should be given to the student in (l) if a different wire gauge is used. The mass per unit length of 26 SWG gauge steel wire is 1.28×10^{-3} kg m^{-1}. Since l_0 is proportional to $1/\sqrt{\mu}$ and hence to $1/d$, the value of $l_0 = 0.5$ m can be scaled appropriately. Alternatively, values for μ can be looked up in Table A3.2 of the student text.

7G STATIONARY WAVES ON A WIRE (II)

This experiment is considered preferable to experiment 7F if you wish to use only one sonometer experiment. Firstly, the frequency f can be varied continuously using a fixed length of wire l. With tuning forks it is really only possible to vary l so as to obtain resonance with a particular tuning fork. Secondly, the resonance is much easier to detect, since the applied force is significantly larger than can be obtained using a tuning fork. Finally, it is possible to investigate the harmonics on the wire: they are easier to see (or hear) and with tuning forks you are limited at best to the first two harmonics (256 and 512 Hz). The precision of the final results may not be as good as using tuning forks because the scales of signal generators can be incorrectly calibrated.

As in experiment 7F the maximum tension should be reduced if a thinner wire than $d = 0.46$ mm is used. However, only if the diameter is a factor of two larger need you tell students to start below 100 Hz on the signal generator. For thinner wires, the first harmonic will occur at a higher frequency than given in (l): f is proportional to $1/d$.

The resistance per unit length of the steel wire is about 1.2 Ω m^{-1}, and a current of between 0.5 and 1 A can usually be produced with a standard signal generator on its low impedance output. If a smaller diameter wire is used, the current may be reduced significantly. To offset this problem a stronger magnetic field may be used by adding further magnadur magnets to the yoke. The current should not be increased significantly or else the wire may break even below P_{max}.

8 GEOMETRICAL AND PHYSICAL OPTICS

8.2 SPECIAL VERNIER SCALES

a TRAVELLING MICROSCOPE VERNIER

Many travelling microscopes use the scale shown in Fig. 8.2 in the student's text. Those currently available from suppliers have a different scale, with a main scale graduated in 0.5 mm intervals and a double length vernier reading to 0.01 mm.

8A REFRACTION AT A PLANE BOUNDARY

THE MEASUREMENT OF REFRACTIVE INDEX OF A LIQUID USING THE CRITICAL ANGLE METHOD

A D-shaped block is preferred to a rectangular block. Using the latter requires correct positioning of the reference line. If the refractive index of the liquid is only slightly smaller than that of the block, it is impossible to obtain extinction. The geometrical construction to find θ_c is also simpler using the D-shaped block.

The main problem with water as the liquid is that it evaporates quite rapidly, leaving the paper dry.

MEASUREMENT OF THE REFRACTIVE INDEX OF A LIQUID USING AN AIR CELL

A simple air cell can be made from two glass slides separated by a rectangle of wire and sealed with Plasticine and PVC tape. The cell can then be mounted vertically in a tall 50 ml beaker using a cork with a suitable slot cut to hold the cell firmly. A pointer is fitted to the base of the beaker. See Fig. 4. If the beaker is placed on a sheet of card, its rotation can be measured.

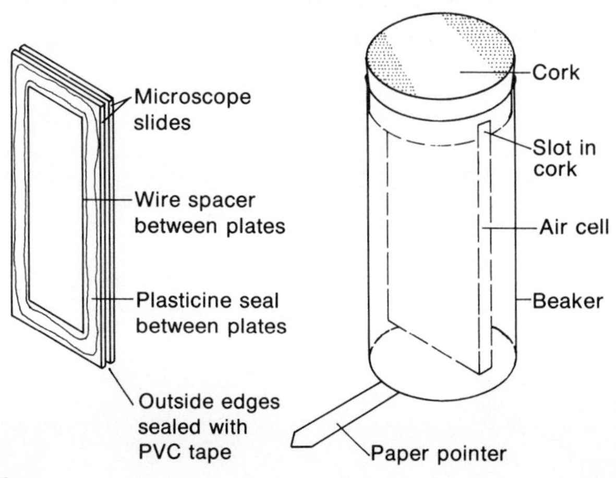

Fig. 4.

8B CONVERGING AND DIVERGING LENSES

For the converging lens a focal length of 0.1 m is preferred. This gives a larger range of u and v for the diverging lens experiment.

W The experiment is more interesting if the students are given an inaccessible lens mounted inside a cardboard tube. d is then measured from one (marked) end of the tube.

8C CONVERGING AND DIVERGING MIRRORS

In order to avoid the problem of using off-axis rays in the measurement of u and v for the converging mirror, students could be given two mounted optical pins, and find the image position by the method of no parallax.

8D DEVIATION BY A PRISM

Other sources of light can be used instead of a sodium lamp. But the sodium (D) orange lines are very easy to see even in subdued daylight.

In the procedure for levelling the table only a broad outline has been given. Often the position of the prism clamp can obscure one of the rays reflected off the face of the prism. This means that the prism cannot be fixed in the position shown in Fig. 8.27 in the student's text, with the face AB perpendicular to the line joining S_1 and S_2. Accordingly the procedure for levelling the table may have to be amended, but the principle is the same.

8E THIN FILM INTERFERENCE

You may wish to omit the sections (e), (f) and (p) to (s), on Newton's rings, since they are not in most A-level syllabuses. However, they are included because students have often found it easier to obtain good interference fringes than with the wedge film. The mathematics of the circular fringe system is more difficult for non-mathematical students. You may wish your students to make qualitative observations on the rings and omit the analysis.

APPARATUS REQUIRED

For the wedge film, glass microscope slides can be used with aluminium foil, such as used for cooking. Several suppliers provide glass plates and suitable materials for making wedge films.

It is possible for students to obtain good fringes using the optical system shown in Fig. 8.33 in the student's text but it is suggested that the reflecting plate could be incorporated in a permanent holder. This holder would also support the wedge or lens on its optical flat. The construction is shown in Fig. 5, with the measurements corresponding to a 0.1 m by 0.1 m glass plate about 2 mm thick.

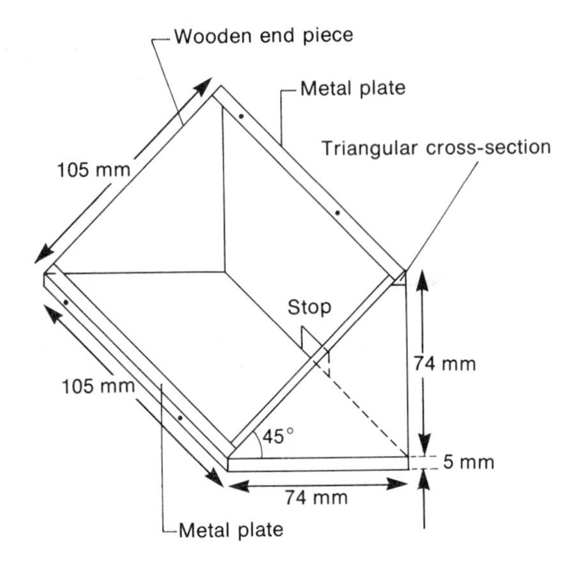

Fig. 5.

The glass plate is supported on a wooden frame made up from a base and two triangular sections. The top corners of the triangular sections are secured with a cross-piece of wood of triangular cross-section. A metal stop is fitted at the back of the base to prevent the glass plates or lens sliding out if the unit is tilted. The glass plate is supported on the top and held in place by two metal strips pinned to the base and the triangular cross-section. Alternatively the glass plate can be held in place by sticking PVC tape around the edges and on to the frame.

Other designs include a frame made up of a base and two square side pieces, with grooves at 45° to hold a glass plate (e.g. Philip Harris P39010/6 supplied with a 32 mm diameter plano–convex lens and optical flat).

The converging lens should have surfaces with a radius of curvature of at least 1 m to give a reasonable ring spacing. For ease of manipulation a 64 mm diameter lens is recommended.

s The measurement of *R* could be checked using a spherometer. This was not included as the experiment was considered to be quite long as written, without including additional material on the use of a spherometer.

8F DIFFRACTION GRATING SPECTRA

The replica diffraction grating should not be blazed otherwise students will only see one diffraction order.

If one of the sources listed is not available, other sources can be used, e.g. cadmium vapour with wavelengths of 467.8 nm (blue), 480.0 nm (blue), 508.6 nm (green) and 643.8 nm (red).

i The levelling procedure has intentionally been left simple. The standard procedure is quite lengthy and it varies according to the positioning of the grating holder with respect to the levelling screws. The procedure given does work and enables the student to get on with the main purpose of the experiment – measuring the spectral lines.

r Rather than just produce a set of wavelengths for the mercury spectrum, we had considered relating these wavelengths to the electronic transitions in the energy level diagram for mercury. However, the energy level diagram consists of singlet (*S*) and triplet (*T*) states and transitions are subject to a set of selection rules. This was considered too complicated for analysis at A level.

9 THERMAL PROPERTIES OF MATTER

9A CALORIMETRY

Other materials can be used if aluminium blocks are not available. If brass ($c = 370$ J kg^{-1} K^{-1}) or iron ($c = 480$ J kg^{-1} K^{-1}) are used, the time of heating can be reduced to about 10 minutes, and temperature readings should be taken every half minute.

If no hair dryer is available the experiment should be completed assuming that Newton's law of cooling is valid in still air. The error in the temperature correction as a result of this assumption is usually less than 2 K.

9B THERMAL CONDUCTIVITY (I)

Some old versions of Searle's bar use steam heating instead of electrical heating. This means that only a single value of λ can be determined and no indication of the energy loss along the bar can be obtained. This experiment is of doubtful merit even in the form described, since it involves a large amount of waiting and very few measurements. Without the possibility of determining two values for λ, this experiment is considered to be suited only for demonstration. 0–50 °C thermometers should suffice for measuring the temperatures along the bar. If possible, matched pairs of thermometers should be kept for this experiment.

9C THERMAL CONDUCTIVITY (II)

It is important to use as thin a sample as possible, otherwise the samples do not reach a steady state in a normal practical session. For glass

$$\lambda = 0.8\text{–}1.3\,\mathrm{W\,m^{-1}\,K^{-1}}$$

the sample can be up to 3 mm thick; for Perspex and cardboard (both have λ values of about 0.2 W m^{-1} K^{-1}) the sample should be 1 mm or less in thickness. Cork ($\lambda = 0.04$–0.05 W m^{-1} K^{-1}) is not recommended. Perspex does warp after heating and samples may have to be replaced frequently.

9D SATURATED VAPOUR PRESSURE

A capillary tube of bore between 0.5 mm and 1 mm and about 15 cm length is sealed at one end. The index is introduced into the capillary by heating the tube along its length and then immersing the open end into a beaker of water (coloured with potassium permanganate). It is a matter of trial and error to produce an index about 10 mm long and an enclosed air column about 20–25 mm in length as the air cools. The length of the air column increases rapidly as the boiling point is approached: the length increases by a factor of about 4 from room temperature to 90 °C and a factor of about 9 from room temperature to 95 °C. Care must be taken therefore not to heat the air column above a temperature that will cause the index to be pushed out of the tube. The shorter the length of the air column at room temperature, the higher the temperature that can be used. This does mean that measurements at low temperatures are not very accurate.

To reduce evaporation of the index during storage, a piece of rubber tubing is fitted to the open end of the capillary tube. This is sealed by a Hoffmann or Mohr clip. The use of a piece of rubber tubing sealed by a piece of glass rod has been found to be unsatisfactory. The sudden change of pressure when the rubber tube is removed often fragments the index.

If organic liquids are used, the liquid should be coloured with aniline or an azo dye.

9E LATENT HEAT OF VAPORISATION

Ethanol is the preferred liquid since it has a far lower specific latent heat of vaporisation than water (0.84 MJ kg^{-1} compared with 2.26 MJ kg^{-1}) and a lower boiling point than water (79 °C compared with 100 °C). It therefore takes a far shorter time to collect the required quantities of condensed vapour. Unfortunately pure ethanol is expensive. Methylated spirits cannot be used instead, because it contains methanol as an impurity. Methanol has a lower boiling point (64 °C) than ethanol; so in any experiment the methanol will distil over first. Since the value of L_v for methanol is significantly larger than the value for ethanol (1.12 MJ kg^{-1} compared with 0.84 MJ kg^{-1}), two successive experiments will produce significantly different values for L_v.

If distilled water is used, it can be pre-boiled so as to save time in heating the water in the apparatus from room temperature. It will still take about 30 minutes to collect 50 cm^3 of water. Collecting only 25 cm^3 is not recommended if a correction is to be applied for energy losses. The percentage error in the mass difference $(m_1 - m_2)$ could exceed 5%, far larger than the error expected from a neglect of energy losses.

10 CURRENT ELECTRICITY

10A DIODE CHARACTERISTICS

This version of the experiment was preferred to the standard type of experiment on the characteristics of various components, e.g. a resistor, a filament lamp and different types of diode. Emphasis is placed on obtaining the forward characteristics of a silicon and a germanium junction diode for small current values. Under these conditions the data obtained is suitable for data analysis using the 'rectifier equation'.

Suitable diodes are 1N4001 for silicon and OA47 for germanium. A resistor of 1200 Ω limits the current in the diode to about 2 mA. Significant deviations from the 'rectifier equation' occur for currents in excess of this value.

A 2 V stabilised d.c. power supply may be used instead of an accumulator; the variable resistor (rheostat) used as a potential divider is retained to give the fine adjustments in V.

10B METER CONNECTIONS

Although this is quite a long experiment, it is considered a valuable exercise in helping students to understand how meters (of the moving coil type) affect the value of a resistance calculated from V/I, when neither of the meters is ideal: the voltmeter has a fairly low resistance (10 kΩ) and the ammeter has a significant resistance (100 Ω). Of course, the increased use of digital meters, which are near to ideal voltmeters and ammeters, does place a limit on the value of this experiment.

The experiment is designed for use with Unilab grey meters with a resistance of $R_g = 1000\ \Omega$, a full-scale deflection (f.s.d.) corresponding to a current $I_g = 100\ \mu A$ and a potential difference of $V_g = 100$ mV. A shunt resistance of $R_g/9 = 111\ \Omega$ in parallel with the basic meter is used to produce an ammeter with a f.s.d. of $10I_g = 1$ mA. A multiplier of resistance $9R_g = 9000\ \Omega$ in series with the basic meter is used to produce a voltmeter with a f.s.d. of $10V_g = 1$ V.

Other meters can be adapted using the same ratios: $I = 10I_g$ with the resistance of the ammeter $R_a = R_g/10$, and $V = 10V_g$ with the resistance of the voltmeter $R_v = 10R_g$.

The analysis leading to the value of R, for which both circuits X and Y of Fig. 10.12 in the student's text give the same percentage errors in R, is given here. You may calculate the appropriate value for R and its percentage error should the factors used not be exactly those just given.

(i) In circuit X, the voltmeter reads correctly, whilst the ammeter reading is too large. The calculated value of R, R_c is then given by the value of R and R_v in parallel and the error in R is:

$$R_c - R = \frac{R R_v}{(R+R_v)} - \frac{R(R+R_v)}{(R+R_v)} = (-)\frac{R^2}{(R+R_v)}$$

The negative sign can be ignored, since we are only interested in the magnitude of the error.

(ii) In circuit Y, the ammeter reads correctly, but the voltmeter reading is too large.

The calculated value of R, R_c is $(R + R_a)$ and the error in R is R_a.

The errors in R from (i) and (ii) are equal when:

$$R_a = \frac{R^2}{(R+R_v)}$$

or:

$$R = \frac{R_a + \sqrt{(R_a^2 + 4R_aR_v)}}{2}$$

since only the positive root of the quadratic is physically valid. If $R_a = R_g/k$ and $R_v = kR_g$, then:

$$R = \frac{R_g}{2}\left[\frac{1}{k} + \sqrt{\left(\frac{1}{k^2} + 4\right)}\right]$$

The percentage error in R is then:

$$100\frac{R_a}{R} = \frac{200}{[1+\sqrt{(1+4k^2)}]}$$

For the factor $k = 10$ chosen in this experiment: $R = 1051\ \Omega$ and the percentage error in R is 9.5%. Clearly, the largest error in R of 61.8% occurs when unadapted meters ($k = 1$) are used and the error decreases as k is increased.

This analysis assumes that there are no errors in the meters. If either or both of the meters reads incorrectly as a result of an incorrect shunt or multiplier, then the value of R for which both circuits X and Y give the same error will differ from the value calculated above. If there is a 2% error in the ammeter reading, such that it reads 1 mA when the actual current is 0.98 mA (e.g. $R_a = 102$ instead of 100), then the percentage errors for R in circuits X and Y are equal for $R = 860\ \Omega$. If there is a 2% error in the voltmeter reading, such that when it reads 1 V the actual voltage is 0.98 V (e.g. $R_v = 9800\ \Omega$ instead of 10 000 Ω), then the percentage errors in R in circuits X and Y are the same for $R = 1290\ \Omega$. However, this wide range of values for R does not affect the general shape of the two error curves for circuits X and Y.

10C ENERGY TRANSFER

A suitable source of e.m.f. consists of 3 1.5 V dry cells mounted in a 6 V battery unit, with the artificial internal resistance of 47 Ω mounted inside a hollow cylinder in the fourth cell position. Alternatively, the internal resistance may simply be soldered across the two contacts of the fourth position in the battery unit. Without such an added internal resistance, it is very difficult to obtain power and efficiency curves suitable for analysis.

As an alternative to the 100 mA ammeter, a digital meter used on its 200 mA setting is suitable.

GENERAL POINTS FOR POTENTIOMETER AND METRE BRIDGE EXPERIMENTS

WIRE

1 m of 26 SWG nichrome wire with a resistance of 6.6 Ω is recommended; the metric equivalent is wire of diameter 0.4 mm with a resistance of 8.6 Ω. This limits the current in the potentiometer/metre bridge wire to less than 0.5 A and avoids both significant electrical heating of the wire and falling of the driver cell voltage with time.

CENTRE-READING GALVANOMETERS

For most experiments the centre-reading galvanometers should have an f.s.d. of ±100 μA. In the initial stages of finding a balance point, the current through the meter can be at least ten times this value. The galvanometer must therefore be protected using a series resistor of at least ten times the galvanometer resistance. This resistance can be shorted out by connecting a push button switch in parallel with the resistor. For a galvanometer resistance of 1750 Ω, a convenient series resistor is 22 kΩ, limiting the current to less than 100 μA.

If a normal centre-reading galvanometer is not available, then a 100 μA meter may be used, with its pointer reset to the centre using the screw adjustment usually used to zero the meter. Then the series resistor should be increased to 47 kΩ.

For potentiometer calibration experiments using a standard cell (experiments 10D, 10F and 10G) and in the measurement of small e.m.f.s (experiment 10G), the current through the galvanometer will be less than 10 μA. Therefore, a light-beam galvanometer with a maximum sensitivity of about 20 mm μA^{-1} will be required. In calibration experiments the galvanometer will usually be used on its ×0.01 or ×0.1 scales; for the measurement of the thermo-electric e.m.f. it will be used on its ×0.1 and ×1 scales.

DRIVER CELL

A 2 V accumulator is normally used, although there is no reason why a stabilised d.c. supply should not be used. However, care must be taken to limit the setting to about 2 V. Exceeding this value by even 2 V can cause significant heating of the wire.

STANDARD RESISTORS

The following are required: 1 Ω (experiment 10E), 2 Ω (experiment 10F), 5 Ω (experiments 10G, 10H and 10I), 10 Ω (experiments 10E and 10J) and 100 Ω (experiment 10E). Except for the lowest two values, the resistors can be made from resistors with 1% tolerances. The lowest two resistor values dissipate as much as 2 W and so must be wound from a suitable gauge of insulated constantan wire. The resistors are made up as follows:

1 Ω	1.58 m of 20 SWG insulated constantan wire;
2 Ω	3.16 m of 20 SWG insulated constantan wire;
5 Ω	3.36 m of 26 SWG insulated constantan wire;
10 Ω	1.59 m of 30 SWG insulated constantan wire;
100 Ω	1% tolerance wire wound resistor is more convenient than winding a coil.

All resistors should be non-inductively wound by folding the wire in two, before it is wound on a former.

STANDARD CELL

The cadmium Weston cell with an e.m.f. of 1.019 V is normally used for calibrating the potentiometer. The maximum recommended current drawn from the cell should not exceed 10 μA. The cell is therefore connected in series with a resistor of 47 kΩ and a resistor of 220 kΩ. The latter resistor can be shorted out with a push button switch connected in parallel. Thus, even if the cell is connected incorrectly in the potentiometer circuit, the current through the cell is unlikely to exceed 10 μA. For absolute safety this resistor could be increased to 470 kΩ, but the sensitivity with which the initial balance point can be found is reduced.

10D POTENTIOMETER (I)

The cell consists of a 1.5 V dry cell connected in series with an added internal resistance of 10 Ω, and mounted on a wooden block with 4 mm terminals. The normal internal resistance of a dry cell is only about 1 Ω and this cannot be measured accurately, because the range of balance lengths is too small except for currents through the cell in excess of 100 mA.

10E POTENTIOMETER (II)

The purpose of this experiment is to allow students to select from a range of standard resistors (1 Ω, 10 Ω and 100 Ω) the most suitable for comparison with a series of four unknown resistors.

The suggested values for the unknown resistors are:

D	2 Ω	0.2 m of 28 SWG nichrome wire or 2.2 Ω (2.5 W) resistor (vitreous);
A	15 Ω	1% tolerance resistors (0.5 W), (metal film)
C	47 Ω	
B	150 Ω	

Wire W is 1.2 m of 28 SWG nichrome wire (9.73 Ω m^{-1}) or 2.4 m of 28 SWG constantan wire (4.41 Ω m^{-1}) with a resistance of about 10 Ω.

The rheostat R_v has a maximum resistance of about 20 Ω, and is used to reduce the potential difference across the wire. This allows the balance points with the various combinations of resistors to be obtained as near as possible to the right-hand end of the wire. The rheostat is not essential as most balance points are between 0.3 and 0.7 m along the wire, provided the correct combinations are chosen for the resistors.

10F POTENTIOMETER (III)

A basic Unilab grey meter (R_g = 1000 Ω, I_g = 100 μA) is adapted to give a f.s.d. of about 1 A using a modified shunt consisting of copper wire. This produces large errors in the ammeter readings at high current values as a result of the electrical heating of the copper shunt and its consequent increase in resistance.

If you wish the systematic error to increase from zero to about 0.06 A at a current of 1 A, then the resistance of the shunt should be 0.1001 Ω. This is made from 68 mm of 40 SWG copper wire or from 106 mm of 38 SWG copper wire. All ammeter readings will then be too high. Alternatively, the shunt resistance can be chosen so that the ammeter reads correctly at about 2/3 of f.s.d., thus giving too small a reading below 0.67 A and too high a reading above 0.67 A. 55 mm of 40 SWG or 86 mm of 38 SWG copper wire should then be used. The standard 2 Ω resistor should be capable of dissipating 2 W without significantly changing its resistance value.

10G POTENTIOMETER (IV)

A copper–constantan thermocouple with an e.m.f. of 42.8 μV K^{-1} is most suitable, giving a maximum p.d. of 5 mV for a temperature difference of 100 K between the junctions.

Students should be given the resistance of the potentiometer wire or it can be measured using a digital multimeter to within ±0.1 Ω.

10H METRE BRIDGE (I)

The two wires used are X: 1 m of 30 SWG nichrome (13.9 Ω m^{-1}) or 2 m of 30 SWG constantan (6.29 Ω m^{-1}); Y: 1 m of 26 SWG nichrome (6.58 Ω m^{-1}) or 2 m of 26 SWG constantan (2.53 Ω m^{-1}). A variable resistor (0–20 Ω rheostat) can be included in series with the driver cell so as to limit the current through the metre bridge wire and the wires under investigation to less than 0.1 A. However, this does slightly reduce the sensitivity with which the balance point can be found.

You may like to ask students to investigate the effects of any 'end corrections' on the ratio R/S. This is achieved by simply interchanging R and S and finding the new balance points l_1' and l_2'. Ideally $l_1 l_1'/l_2 l_2' = 1$.

10I METRE BRIDGE (II)

Enamelled copper wire is used for the coil. For a coil with a room temperature resistance of about 4 Ω, increasing to about 6 Ω at 100 °C, 7 m of 36 SWG enamelled copper wire is used. If this is wound on a former of 15 mm diameter, this corresponds to about 150 turns; on a former of 10 mm diameter, this corresponds to about 220 turns. The coil should be non-inductively wound, by folding the wire in half before it is wound on the former. The final coil is placed in a test tube filled with glycerol. The ends of the coil can be stripped of enamel and soldered to low resistance wires for connection to the metre bridge. Ideally, the test tube should be fitted with a rubber bung and the wires taken through two holes in the bung. This avoids the risk of spilling the glycerol or accidentally unravelling the coil.

Provided that the current in the copper coil is less than about 0.2 A, the effects of electrical heating of the coil can usually be neglected. If there are doubts about this effect, the measurements can be made with a reduced current, by connecting a 0–20 Ω rheostat in series with the driver cell of the metre bridge.

10J METRE BRIDGE (III)

The most suitable thermistor is TH-7, with a resistance of about 90 Ω at 0 °C, 40 Ω at 20 °C and 4 Ω at 100 °C. The thermistor should be soldered to two low resistance wires for connection to the metre bridge. It is advisable to make these wires at least 0.5 m long, so that they can be trailed over a clamp set on its stand, so as to keep the wires clear of the Bunsen flame. Wires trailing over the edge of the beaker in which the thermistor is placed are frequently burnt.

The recommended standard resistor is 10 Ω, representing an average value of the thermistor over the temperature range used. The use of this standard gives balance points ranging from 0.9 m at 0 °C to 0.3 m at 100 °C. Ideally, balance points should be near to the centre of the wire, so that the error in the ratio l_1/l_2 is a minimum; this can be achieved by changing the standard resistors as the resistance of the thermistor alters significantly, e.g. 100 Ω from 0 to 20 °C, 10 Ω from 20 °C to 80 °C, 5 Ω from 80 °C to 100 °C.

Electrical heating of the thermistor does not seem to be significant provided that the current through the thermistor is less than 0.1 A. This condition is satisfied when the thermistor resistance is larger than about 15 Ω. At 100 °C, the current through the thermistor could be as large as 0.2 A. The simplest way to avoid this problem is to include a variable resistor (0–20 Ω rheostat) and ammeter in series with the driver cell to maintain the total current in the circuit at less than 0.1 A.

11 ELECTRIC AND MAGNETIC FIELDS

11A HALL EFFECT

The permanent magnet may be an eclipse magnet (0.18–0.20 T) or a number of magnadur magnets placed on a yoke; a single pair gives a B value of about 0.03 T increasing up to 0.16 T with three pairs placed inside the yoke. These values can be checked using the search coil and ballistic galvanometer method (see experiment 11B). Note that with a current of 50 mA in the Hall slice, the Hall voltage may exceed 100 mV for the n-type semiconductor; if this is so either the current can be reduced or a 1 V multiplier can be made available.

For the optional experiment a 741 operational amplifier with a gain of -100 should be available with the connections clearly marked for the student.

c We do not recommend the use of a mirror-lamp galvanometer to measure small Hall voltages. Whilst qualitative data can be obtained, we believe that it is quite wrong to allow students to use a low resistance meter to measure potential differences. For a typical galvanometer of this type an error of a factor of 8–10 can occur in measuring a potential difference of 1 mV.

11B BALLISTIC GALVANOMETER

I COMPARISON OF CAPACITORS

A suitable value for the unknown capacitor is 0.47 μF. A larger value may give a deflection off the scale of the galvanometer, unless the pointer is zeroed on the extreme left-hand side of the scale.

II USE OF THE SEARCH COIL TO MEASURE MAGNETIC FIELDS

We thought that it would be useful for students to make the choice of the resistance value to be included in the search coil circuit. Mounted resistors with nominal values of 15 kΩ, 33 kΩ, 68 kΩ and 100 kΩ should be sufficient to give a reasonable choice for B values in the range 0.03–0.20 T. The charge sensitivity of the galvanometer has been assumed to be about $80\,\text{mm}\,\mu\text{C}^{-1}$. The search coil parameters are $N = 5000$, $A = 10^{-4}\,\text{m}^2$ and $R_s = 2000\,\Omega$.

11C ALTERNATING MAGNETIC FIELDS

A 'slinky' spring should be reserved for this experiment. Often a 'slinky' used to demonstrate waves has a large number of kinks, and it is unsuitable for this experiment, where a constant value of the number of turns per metre is desirable.

11D ELECTROMAGNETIC INDUCTION

Although the detailed behaviour of the transformer is outside the A-level physics syllabus, we thought that it would form the basis of a novel investigation. It seemed a good idea to give students the opportunity to find out how the flux linkage between the primary and secondary coils alters the secondary voltage. Also, the dependence of the secondary voltage on the frequency of the primary voltage supply may be a surprise to students familiar with the standard equation $(V_s/V_p) = (N_s/N_p)$ for a transformer. Our explanation of this variation given in the student guide under (j) has not been confirmed.

Suitable coils for the experiment are Unilab coils with 60 and 120 turn tappings.

p Clearly an inverse law cannot be valid at small values for the separation x of the coils, but it was surprising that the relationship was not valid for the larger values of x.

q We have no explanation for the success of the exponential relationship in describing the variation of V_s with x. We did find that this relationship produced linear plots for a range of coils and materials (e.g. Mu-metal) used as transformer cores.

11E SPECIFIC CHARGE OF AN ELECTRON (I)

We recommend that the fine beam tube should be mounted ready for the students. It is relatively easy to break the plastic cap–glass seal. The filament supply should be fixed at 6.3 V. Students (and teachers) have been tempted to increase the filament voltage above this value, if they have difficulty in seeing the electron beam. Above about 9 V damage is likely to occur as a result of the vaporisation of the cathode.

11F SPECIFIC CHARGE OF AN ELECTRON (II)

We have listed only one 0–5 kV power supply providing both the accelerating voltage and the deflecting field. Clearly this limits the flexibility of measurements on the deflection of electrons in an electric field. The deflection will be independent of V_A, since E is increased proportionally. If a second 0–5 kV supply is available, students can determine a series of e/m_e values for varying electron speeds.

Note that we have found that some HT supplies incorporating a voltmeter gave readings that were as much as 50% too low. This leads to e/m_e values that are as much as a factor of two in error. For this reason we recommend the use of a separate, high impedance voltmeter to measure V_A and V_d. An avometer used on its d.c. range (setting 1000 V) with terminal connections 2500 V D.C.− and D.C.+ has been found to give more reliable readings. The input impedance of the avometer on this range exceeds 10 MΩ.

12 CAPACITANCE

12A DIRECT MEASUREMENT OF CAPACITANCE

a The standard mirror-lamp galvanometer has a resistance of about 14 Ω only, so the accidental connection of the power supply across the galvanometer may result in permanent damage to the meter. A 100 kΩ resistor limits the maximum current to 200 μA. *CR* may not have negligible effect for $f = 500$ Hz, depending on the precise geometry used for the parallel plate capacitor.

c A 100–200 mA fuse can be included in the circuit between the d.c. voltage supply and the vibrating reed switch to protect the reed switch from accidental short-circuiting of the capacitor plates.

Further work

u–w This addition was considered to be useful for introducing the students to measurement of ε_r for liquids. If you do not wish to make a separate parallel plate capacitor unit, then two plates (one covered with polythene) can be immersed in a large tank of water and polythene spacers used to maintain the required gap for the water between the plates. Significantly smaller plates than 0.25 m × 0.25 m can be used, since the capacitance is increased by a factor of about 80 and *A* can be reduced by a factor of 10 or more if required.

12B CHARGING AND DISCHARGING A CAPACITOR (I)

Recommended values: known capacitor $C = 2200$ μF electrolytic; voltmeter resistance $R = 50$ kΩ (100 μA f.s.d., $R_g = 1000$ Ω) giving a f.s.d. of 5 V for a Unilab grey meter. $RC = 110$ s.

Other voltmeters may be used and the value of *C* chosen to give an *RC* value of about 100 s. Unknown capacitor $C = 1000$ μF electrolytic. One frequent problem is that after an initial fall in V_R, the reading of the voltmeter remains almost unchanged. This is usually a result of a damaged electrolytic capacitor. You should tell students that if V_R has not decreased to $\frac{1}{10}$ of its initial value in 5 minutes, it is likely that the capacitor requires replacement.

12C CHARGING AND DISCHARGING A CAPACITOR (II)

Normally, the maximum current in the *RC* circuit is significantly less than the maximum current for the reed switch (about 250 mA), since $R > 100$ Ω. Should *R* be accidently set well below 100 Ω, the current could be high enough to damage the reed switch contacts. Therefore, as a precaution, a 100–200 mA fuse can be included in the circuit between the d.c. voltage supply and the vibrating reed switch.

13 ALTERNATING CURRENTS

Digital meters are recommended for measuring V and I in all a.c. circuits. Some moving coil meters have a significant inductive effect at high frequencies ($f \approx 500$ Hz). This leads to incorrect values for V and I. This is particularly noticeable in experiments 13B, 13C and 13D. In experiment 13B the impedance of an inductor is measured and analysed to give values for L and R; these can be in error by at least 10% if moving coil meters are used. In the series (experiment 13C) and parallel (experiment 13D) resonance circuits we have found that the measured resonant frequency is significantly different (by 10–20 Hz) from the theoretical value based on the values for L, C and R.

13A CAPACITIVE REACTANCE

A standard 1 μF capacitor and a standard 1000 Ω (non-inductively wound) resistor are preferred, since students then obtain an excellent agreement between theory and experiment.

13B INDUCTIVE REACTANCE

A Unilab 100 mH inductor has a resistance of about 55 Ω.

If certain moving coil meters are used for the ammeter and the voltmeter, the results for Z become increasingly in error as f increases. The errors become significant for $f > 500$ Hz, increasing from a 10% error in Z at 500 Hz to a 25% error at 1000 Hz. In these circumstances it is advisable to restrict measurement to a maximum frequency of 500 Hz, otherwise there is unlikely to be a linear section of the Z–f graph at high frequencies.

Our measurements gave values for L as low as 80 mH using certain moving coil meters.

13C SERIES RESONANCE

To obtain good agreement between the experimental results and the theoretical result for f_0 it is important to use a standard 1 μF capacitor and a non-inductively wound resistance box for R. Using digital meters is less critical here than in experiment 13B, because the resonance curves will be internally consistent. However, the resonant frequency may be too high by 10–20 Hz if moving-coil meters with a significant inductive component are used.

There will be different views as to whether we should have included the mention of both charge and current resonance in electrical circuits and the comparison with mechanical systems. Observant students may, however, have noticed the differences between the amplitude resonance curves presented under s.h.m. and the current resonance curves presented separately under a.c. in most A-level text books. It is for this reason that we chose to bring the students' attention to the comparison that is often incorrectly made.

13D PARALLEL RESONANCE

Comments on components and meters are as for experiment 13C.

13E LISSAJOUS' FIGURES

MEASURING THE PHASE ANGLES IN A SERIES C-R CIRCUIT

The 1 μF capacitor should be a standard and the resistor (or resistance box) should be non-inductively wound. Otherwise the agreement between the experimental and theoretical values of ϕ will be poor, particularly at high frequency when the inductive reactance $X_L = 2\pi fL$ becomes significant.

For ease of obtaining Lissajous' figures of the correct proportions, a CRO with variable sensitivities on both the Y- and X-plates is preferred.

CALIBRATION OF A SIGNAL GENERATOR

Most usually, differences between the marked frequency and the measured frequency are a result of the difficulties in reading the scale on the signal generator. Occasionally a signal generator does have a systematic error. We found it made the experiment more interesting if such a signal generator is selected. To do this we used a frequency meter to check each calibration of the signal generator.

14 ELECTRONS, ATOMS AND NUCLEI

14A THE PHOTOELECTRIC EFFECT

Unfortunately, this experiment for verifying the Einstein equation for photoelectric emission and for the determination of h/e has been largely discredited by R.G. Keesing of the Physics Department, University of York. His main point is that there cannot be a well defined cut-off potential V_s, corresponding to the maximum energy of the photoelectrons for a given photon energy hf, except at absolute zero. Only at 0 K would the work function $e\phi$ define precisely the minimum energy required to liberate an electron from the free electron gas in a metal. At room temperature the energy required to liberate an electron from the surface of a metal varies from $e\phi$ down to very small values for the most energetic electrons. Of course, there are relatively few electrons with thermal energies approaching 2 eV, but it is no longer expected that a precise value for V_s will be obtained for the cut-off of photoelectric emission. Also, photoemission can occur for a photon energy hf that is less than $e\phi$. In his paper on the photoelectric effect (*European Journal of Physics* **2**, 139–149 (1981)), Dr Keesing demonstrates not only the absence of a sharp cut-off in V_s but also a dependence of the maximum energy of the photoelectrons on the intensity of the light. It is even possible to obtain a larger value for V_s for an intense red line than for a weak blue line; this would lead to a negative value for h/e.

These results do indicate a flaw in the original equation for the photoelectric effect except at absolute zero. In practice, the detection sensitivity of the apparatus used in this experiment was insufficient to show a significant tail in the photoelectric signal as V_s is increased for a given frequency of the light. There was evidence for small variations in V_s (± 0.2 V) depending on the intensity of the light incident on the cathode, but this could just as easily have been attributed to variations in the sensitivity of the 'balance point' as the intensity of the light varies.

In view of the dependence of the results on the light intensity, this experiment cannot be considered as an accurate method for the measurement of h/e. However, despite variations in the intensities of the spectral lines of the mercury source, internally consistent results have been obtained and a negative value for h/e has never been obtained. Neither has a significant photoemission been detected for red light ($\lambda = 620$–700 nm).

e The use of white light and filters with a fairly large wavelength transmission band is not recommended. Students find it difficult to analyse the results and the value for h is usually a factor or 1.5–2.0 too low unless the correct straight line is fortuitously drawn.

m Alternative filters (single wavelengths) are:

Leybold 46833	404.7 nm
Kodak Wratten W50	435.8 nm
Kodak Wratten W74	546.1 nm
Kodak Wratten W73	577.0/579.0 nm

The value of V_s for the longest wavelength is likely to be near to zero ($\lambda_0 \approx 560$ nm, theoretically, for a potassium cathode).

t The most logical argument for using λ_{min} (f_{max}) of the bands given is that this corresponds to the maximum value of V_s measurement. This reason fails because at the extreme ends of the band the transmission of the filter can be as low as 10% of the maximum transmission. So the effect of λ_{min} on the measurement of V_s can be quite small. The maximum contribution is likely to come from the centre of the ranges given, with some bias towards the low wavelength end of the band. It is virtually impossible to decide on which straight line to draw.

y Examining photoemission for $hf < W$ will require the battery polarity to be reversed in the photoelectric unit. This will involve unscrewing the lid on the unit. Care must be taken in removing this front panel. The wires between the panel and the battery holder are quite short. The surface of the photocell must not be touched.

14B BOHR THEORY

If dark-room facilities are not available, then a large black box with a slit or with a curtain at the front may be used. All the preliminary alignments of the spectrometer and the hydrogen discharge tube may be achieved in subdued light conditions.

14C RADIOACTIVITY

RADIOACTIVE SOURCES

$^{226}_{88}$Ra is not a β-emitter but two of the subsequent radioactive nuclides, $^{214}_{82}$Pb and $^{214}_{83}$Bi, are.

There are no alternative radioactive sources with as high an activity as $^{226}_{88}$Ra for β- and γ-emission: $^{60}_{27}$Co is a β-, γ-emitter but it is often supplied with an aluminium cover which prevents a significant number of β-particles leaving the source. $^{239}_{94}$Pu and $^{241}_{95}$Am are α-, γ-emitters but the γ-radiation has a low energy (< 0.1 MeV), insufficient to penetrate more than about 0.1 mm of lead.

$^{90}_{38}$Sr is a β-emitter only.

DETECTORS

The standard Geiger–Müller tubes are Mullard ZP 1481 (MX168) which detects α-, β- and γ-radiation and the Mullard ZP 1530 (MX142) which detects β- and γ-radiation.

ABSORBERS

Standard sets of aluminium and lead sheets are available: aluminium, set of 7 $\rho x \approx 0.27$–8.90 kg m^{-2} (thickness about 0.1 mm to 3 mm); and lead, set of 4 $\rho x \approx 18$–72 kg m^{-2} (thickness about 1.5 mm to 6 mm). Aluminium sheets to $\rho x = 17.20$ kg m^{-2} and lead sheets to at least $\rho x = 108$ kg m^{-2} can be built up.

a This experiment has been oversimplified. Most radioactive nuclides in the $^{226}_{88}$Ra series emit one or more γ-rays. The equation for absorption strictly applies to a monoenergetic γ-ray. The student will therefore find an average value for the mass absorption coefficient, weighted towards the high energy γ-radiation emitted.

c Despite the emission of at least two different β-particles and several γ-rays from the $^{226}_{88}$Ra series, the distinction between the β- and γ-radiation absorption is quite well defined.

14D RADIOACTIVE DECAY

You should, whenever possible, check the internal 9 V battery of the d.c. amplifier in advance. If accidentaly left on for a day the battery voltage may fall significantly below 9 V and students will not obtain any results. The recharging of the Unilab battery takes several hours.

INTRODUCTION TO ELECTRONICS

Most modern A-level courses now include a substantial amount of electronics. The emphasis has recently moved away from the study of transistor characteristics, and transistor amplifying and switching circuits, towards the use of analogue and digital integrated circuits. In preparing material for these two electronics sections the authors have borne this trend in mind. It is intended that all the experimental work in these sections is performed with integrated circuits.

The opportunity exists for students, even those following a traditional syllabus, to cover electronics mainly by experimental work. Indeed, several examining boards have expressed the hope that their electronics sections will be covered in this way. We envisage students spending every lesson for two or three weeks in the laboratory, engaged in more or less one continuous experimental investigation.

For this reason, and because the treatment of electronics with integrated circuits in nearly all existing A-level physics texts is hardly adequate, to say the least, we have intentionally extended these two sections in the main text. They both contain far more theory and explanatory notes than the other sections in the book. In fact, we have attempted to present to students all the necessary theory and background information so that they will be able to work with the minimum of interruption.

For each section, a student should be provided with a basic experimental kit, and, with a few extra components required here and there, the student can then work through all the experiments or parts of experiments that the teacher selects. No attempt has been made to standardise the length of the experiments to fit the traditional double-period laboratory session. Each experiment deals with one particular topic or function that electronics can perform, but there are one or two instances of an important idea that is developed through several experiments, and students should be encouraged to regard each section as one extended investigation. If the experimental kit is well designed and easy to assemble, work can continue profitably during single periods.

Most of the general information that teachers might require may be found in the main text at the beginning of each section. There are no mystery components to prepare, details of which might need to be kept from students. In the pages that follow some guidance is given to those teachers who would prefer to construct their own apparatus. Many proprietary experimental systems are available, and they all have their advantages and disadvantages. Most would be perfectly adequate for these experiments. However, it is the authors' opinion that a home-made system based on solderless prototyping breadboard is hard to beat (in terms of both cost, convenience and versatility in use), providing time can be found for its construction.

Two types of unit or 'module' can be built, one for each section. These would constitute the 'circuit assembly board or system' that appears at the head of every apparatus list in the text. Each type might consist of a breadboard block mounted on the top of a suitable instrument case. All circuit components, including integrated circuits, can be plugged directly into the board and interconnected with single-strand (0.6 mm diameter) PVC-covered wire, which is available in several colours. Connections to power supplies, meters, CROs, signal generators and other equipment can be made from suitable sockets (BNC, 4 mm, etc.) mounted on the sides of the case. Each socket is permanently wired inside to a 1 mm socket mounted on top, beside the board; and single-strand wire with a 1 mm plug on one end is then used to join each socket to a point in the circuit on the board.

If particular sockets on the case sides are assigned for power supply connections, each unit can be fitted with a switch and 'power on' indicator, and permanent power supply connections can be made to the bus rows on the breadboard, which then become the power supply rails. Special facilities can be incorporated into the units as space allows. This would be particularly advantageous for work in section 16; for most experiments in that section the experimental kit can then consist of just a unit, a power supply, connecting leads and a handful of components. Circuits for special facilities that would be particularly useful are given here. They can all be asssembled on one piece of stripboard, housed inside the case.

15 ANALOGUE ELECTRONICS

All the components required are commonly available. The 741 op-amp can be used for all these experiments, and the TL081 is recommended as the 'better alternative' for some of the further work. In experiment 15A the LDR and the thermistor should both have long single-strand wires soldered to their leads, so that they can be connected direct to the circuit board but held and used some distance away. Resistors could be rated at only 0.25 W, but 0.5 W ones are bigger and rather easier to manipulate; a list of all those required is given in section 15.4.5. Capacitors can all be non-electrolytic; polyester and ceramic types are available in a wide range of capacitances, and with a working voltage that far exceeds V_s.

POWER SUPPLY

This circuit, based on two complementary ICs, provides a twin regulated ±15 V supply and will deliver up to 100 mA per rail. The whole circuit should be housed safely in a suitably robust case, fitted with mains lead and plug, and clearly identified output terminals. See Fig. 6.

BREADBOARD UNIT

Three bus rails will be required (one each for $+V_s$, ground, and $-V_s$). Most solderless breadboard blocks have only two, but extender strips are available that can be attached to the main block. Alternatively, one column of five interconnected sockets on the board can be used for ground. The following extra facilities would be useful and might be incorporated into the unit. Note that for the current limiting resistors in series with indicator LEDs, the resistances shown are suitable for a power supply voltage $V_s = 15$ V. If a different value of V_s is used, these resistances should be altered accordingly. You may also have to alter resistance values to obtain a more suitable brightness, depending on the type and colour of each LED.

TWIN POWER SUPPLY SWITCH AND INDICATORS

If this is fitted it can be used to isolate the board when alterations to the circuit are made. The mains switch on the power supply can then be left alone; it is not as suitable for isolating the board because the smoothing capacitors tend to maintain the outputs for a few seconds after the mains has been isolated. Each rail has its own indicator, which will not light if that half of the power supply fails, or if the rail is accidentaly connected to ground in the circuit under investigation. The protective diodes, which are optional additions, will prevent damage to op-amp ICs should the unit be connected to the power supply with the wrong polarity. Their inclusion will cause a small drop in rail voltage. See Fig. 7.

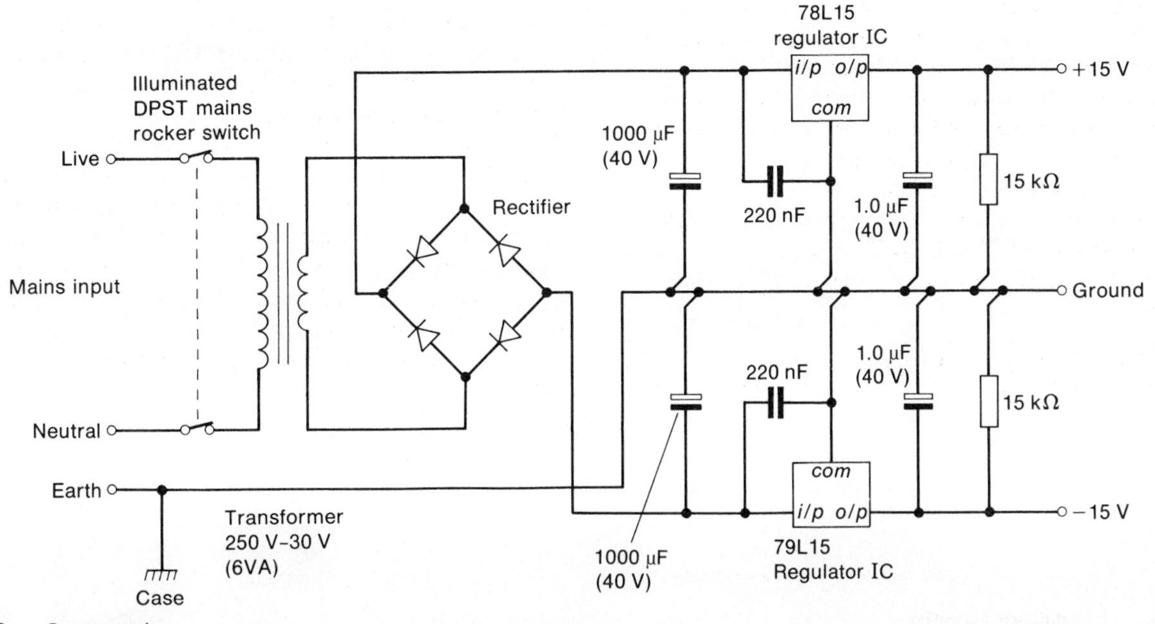

Fig. 6. Power supply.

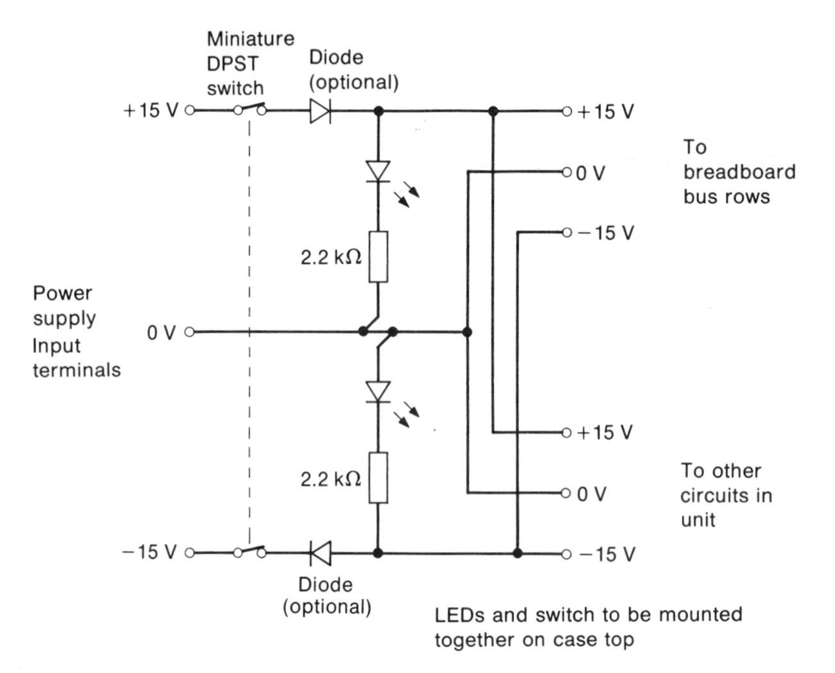

Fig. 7. Twin power supply switch and indicators.

TWO-STATE VOLTAGE INDICATOR

This is a useful device that frees both CRO channels in the switching (comparator) experiments. Each one of the two LED indicators lights only when the voltage applied to the test socket is within 10% or so of one of the two saturation output voltages of the op-amp. One indicates only high voltages (near $+V_s'$) and the other indicates only low voltages (near $-V_s'$). The circuit includes two op-amps, and for convenience use an IC that contains two, like the TL082. See Fig. 8.

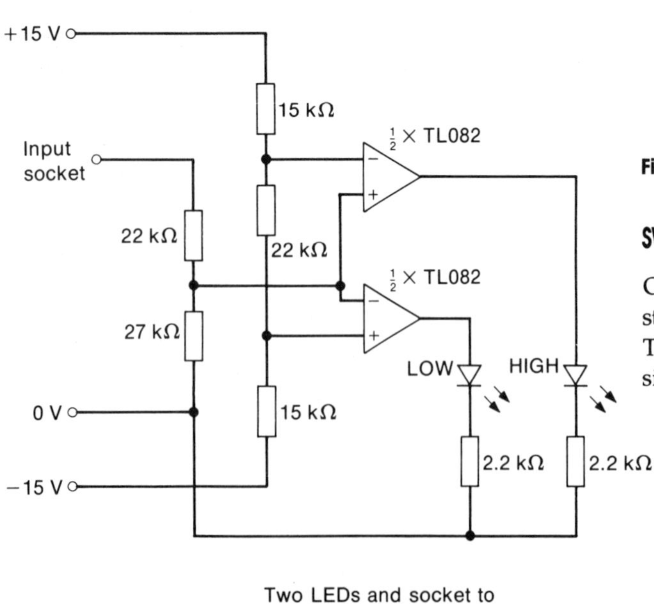

Fig. 8. Two-state voltage indicator.

POTENTIAL DIVIDER

Miniature potentiometers that can be inserted into the breadboard are available, and these would be suitable for the later experiments. However, it is useful to have a couple of potential dividers mounted on the unit to provide constant input voltage levels. See Fig. 9.

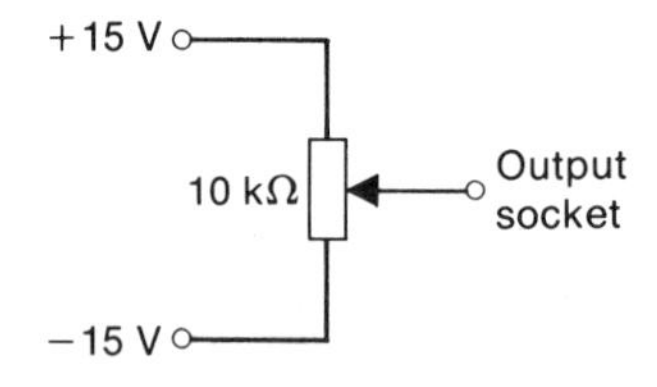

Fig. 9. Potential divider.

SWITCH

One double-pole, single-throw switch will be required to start the integration processes in experiments 15F and 15G. This arrangement allows two batteries to be switched out simultaneously. See Fig. 10.

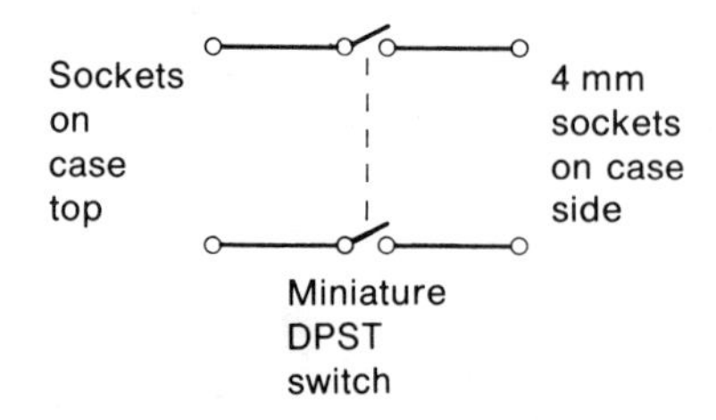

Fig. 10. Switch.

16 DIGITAL ELECTRONICS

A choice has to be made between the two principal logic families: TTL and CMOS. The case for each is presented in section 16.1.6 of the student's text. Overall we would recommend TTL, because in later experiments circuits can then be simplified by leaving many IC inputs open to float to logical 1. But do bear in mind that well-regulated power supplies are essential for TTL circuits, and the extra cost of these might tip the balance in favour of CMOS, especially if the usual low voltage power supplies, which are quite adequate for CMOS circuits if they have some smoothing, are readily available in the laboratory.

Combinational logic only is covered by the first five experiments. The remaining five involve also sequential logic. All the components required are commonly available. Resistors could be rated at only 0.25 W, but 0.5 W ones are bigger and rather easier to manipulate. In experiment 16J the two larger capacitors should obviously be electrolytic, and their working voltage must exceed V_s.

POWER SUPPLY

This circuit (see Fig. 11), based on the 7805 IC, will deliver up to 1 A, more than enough to drive several ICs and LED indicators. It is intended for TTL circuits, but it will, of course, drive CMOS circuits as well. The whole circuit should be housed safely in a suitably robust case, fitted with mains lead and plug, and clearly identified output terminals.

BREADBOARD UNIT

ICs are expensive and easily damaged through careless handling. They are fairly easy to insert into the breadboard but not so easy to remove, and this is when most mishaps will occur. It is worth supplying each student with an extraction tool, the cost of which is about twice that of the average digital IC, and could be justified in one lesson!

The introduction to section 16 contains enough information to enable the student to asssemble LED indicators and logic inputs (including a de-bounced switch) on the breadboard as required. Since, however, these facilities are needed for nearly every experiment, it is well worth incorporating them into the unit. Some suggestions are given here. Note that for the current limiting resistors in series with indicator LEDs, the resistances shown are suitable for a power supply voltage $V_s = 5$ V. If a higher value of V_s is used, with CMOS circuits, these resistances should be increased accordingly. You may also have to alter resistance values to obtain a more suitable brightness, depending on the type and colour of each LED.

POWER SUPPLY SWITCH AND INDICATOR (Fig. 12)

If this is fitted it can be used to isolate the board when alterations to the circuit are made. The mains switch on the power supply can then be left alone; it is not as suitable for isolating the board because the smoothing capacitor tends to maintain the output for a few seconds after the mains has been isolated. The protective diode,

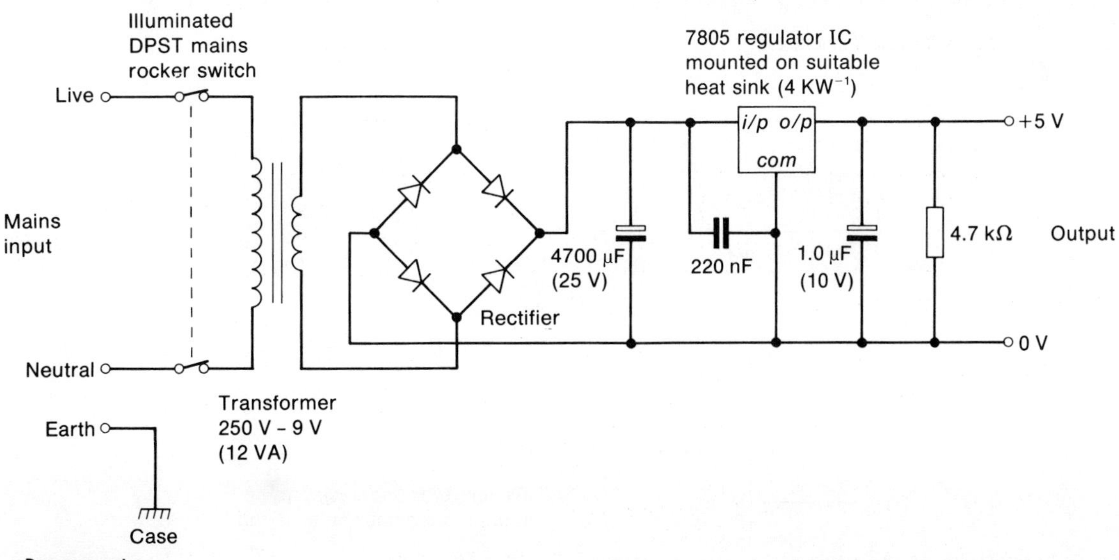

Fig. 11. Power supply.

which is an optional addition, will prevent damage to ICs should the unit be connected to the power supply with the wrong polarity. Note, however, that its inclusion will cause the p.d. between the power supply rails to drop to about 4.3 V, though this will not affect the performance of circuits assembled on the unit.

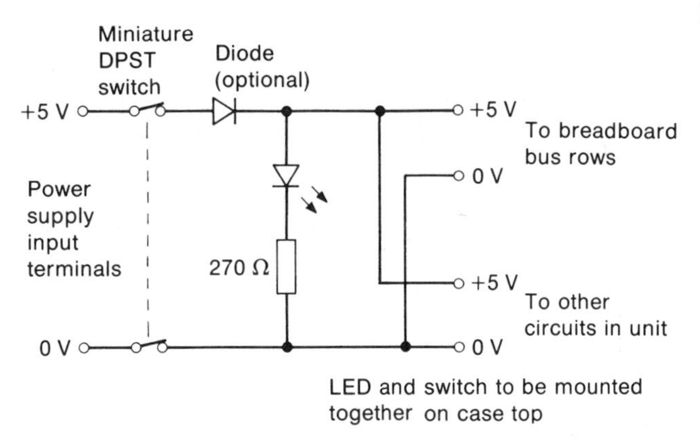

Fig. 12. Power supply switch and indicator.

DE-BOUNCED SWITCHED LOGIC INPUT WITH INDICATOR

This is essential for the later experiments where a clean supply of clock pulses is required. A push switch should be used so that one press/release operation produces one pulse. The output will be at logical 0 with the switch in the position shown in Fig. 14.

OUTPUT LOGIC INDICATOR (Fig. 15)

At least four will be required. The 1 kΩ resistor keeps the input of the inverter at logical 0 when the socket is left unconnected. This arrangement takes advantage of the ability of TTL gate outputs to sink larger currents than they can source. If space permits, one red, one yellow and one green indicator, arranged in a column, would add a touch of realism to the traffic light sequence experiments in 16H.

SWITCHED LOGIC INPUT WITH INDICATOR (Fig. 13)

At least two will be required.

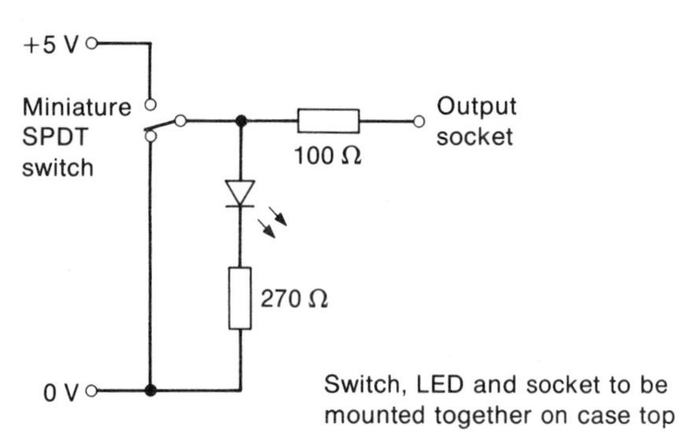

Fig. 13. Switched logic input with indicator.

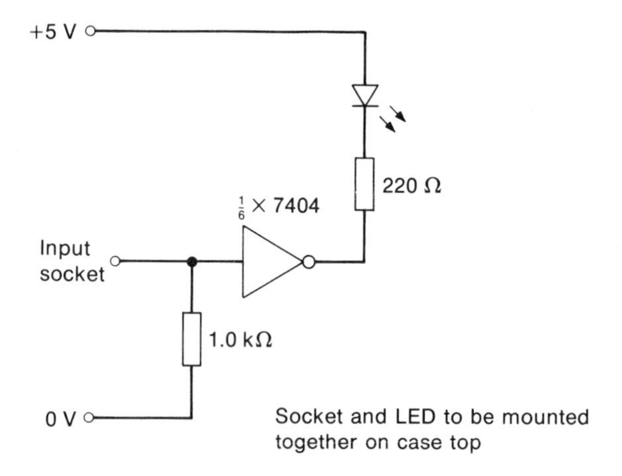

Fig. 15. Output logic indicator.

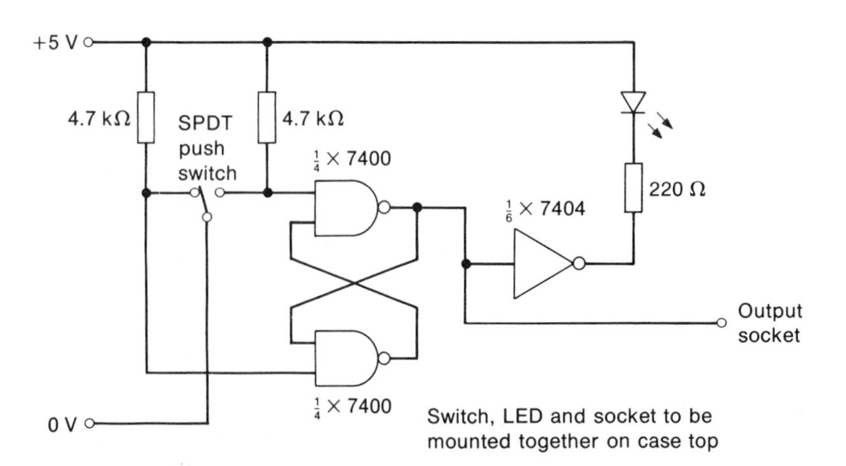

Fig. 14. De-bounced switched logic input with indicator.

PULSE GENERATOR (Fig. 16)

This circuit generates pulses at about 1 Hz, and is intended for use in the later experiments, where a continuous supply of clock pulses is required. The circuit is based on the 555 timer IC, which is available in TTL- or CMOS-compatible form. The indicator, which of course will flash when this circuit is switched on, allows the events that occur in the student's circuit to be related to the beginning or end of each clock pulse.

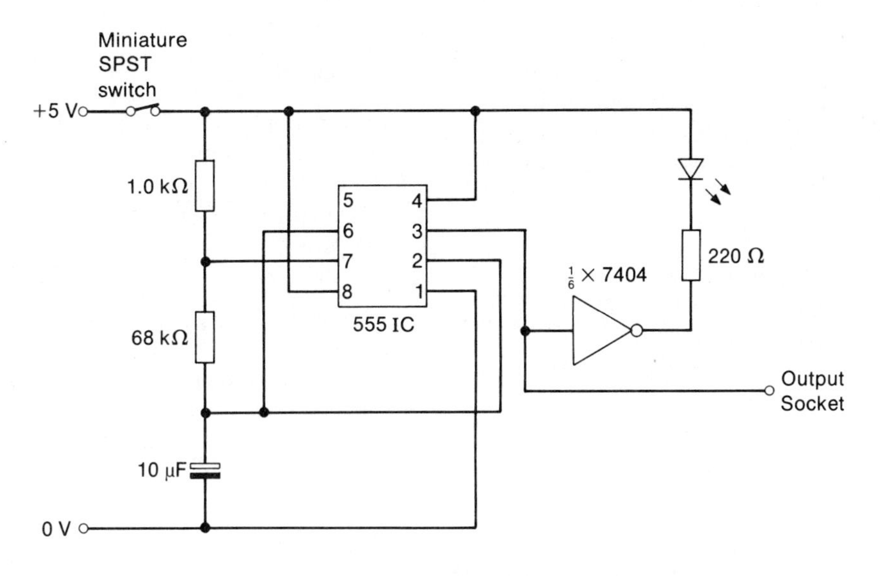

Fig. 16. Pulse generator.

17 PROBLEM EXPERIMENTS

Corresponding to the A-level practical examination questions printed in the student's text, this section reprints the instructions issued by the Examining Boards to schools. The questions are taken from the following Examining Boards:

The Associated Examining Board (AEB)
Joint Matriculation Board (JMB)
Oxford and Cambridge Schools Examination Board (O & C)
University of Cambridge Local Examinations Syndicate (C)
University of London School Examinations Board (L)
University of Oxford Delegacy of Local Examinations (OLE).

The AEB questions are designed to take 45 minutes and the remainder of the questions should take approximately 90 minutes. In the revision of the physics A-level syllabuses, several Examining Boards have taken the opportunity to revise the form of the practical examination: several Boards will be introducing internal assessment as an alternative to the long practical examination.

The problem experiments listed under the following topics may be set as class exercises following the completion of a particular group of experiments or as part of a revision program:

measurement (2 experiments)
mechanical properties of matter (4 experiments)
oscillations and waves (6 experiments)
geometrical and physical optics (8 experiments)
thermal properties of matter (3 experiments)
current electricity (4 experiments)
capacitance (1 experiment)
semiconductor devices (2 experiments)

MEASUREMENT

17A Measurement of the radius of curvature of a concave surface

Candidates are to measure the radius of curvature of two watch glasses (a) by timed oscillations, (b) by geometry, and (c) by optical means.

APPARATUS REQUIRED

- 2 watch glasses 70 mm and 150 mm in diameter; obtainable from Gallenkamp or Griffin
- 1 steel ball of diameter in the range 8 to 12 mm
- Plasticine to support the glasses on the bench
- stopwatch or stopclock measuring to 0.2 s or better
- half-metre or metre rule
- pin, retort stand and clamp

Each candidate will need access to a micrometer screw gauge to measure ball diameter and glass thickness.

Dark screens for section (c) may be provided.

The supervisor must send with the scripts:
the ball diameter
average values of D, h, and t for both sizes of watch glass;
a sample set of all other data.

(O & C 1980)

17B Flow of liquid through a burette

Candidates will be required to determine the rate of flow of water through a capillary tube.

APPARATUS REQUIRED

- 50 ml burette
- capillary tube about 20 cm long and 1 mm bore
- 3 retort stands, clamps and bosses
- 2 300 ml beakers, one of which should be filled with water, covered and allowed to stand overnight
- small funnel
- stopwatch
- metre rule
- plumb bob
- set square
- small piece of rubber or plastic tubing to connect the end of the burette to the capillary tube when they are at right angles to each other
- a piece of thread about 10 cm long

(JMB 1979)

MECHANICAL PROPERTIES OF MATTER

17C Bending of a loaded metre rule

Candidates will be required to investigate the bending of a metre rule.

APPARATUS REQUIRED

- wooden metre rule, as straight as possible, with a small hole drilled in the median section at the 99.0 cm mark
- length of thin string about 5 cm long, knotted at one end and then passed through the hole to the graduated face of the rule and hung over the end of the rule with a small hook on the end – the string can be held in place by Sellotape (see Fig. 17).

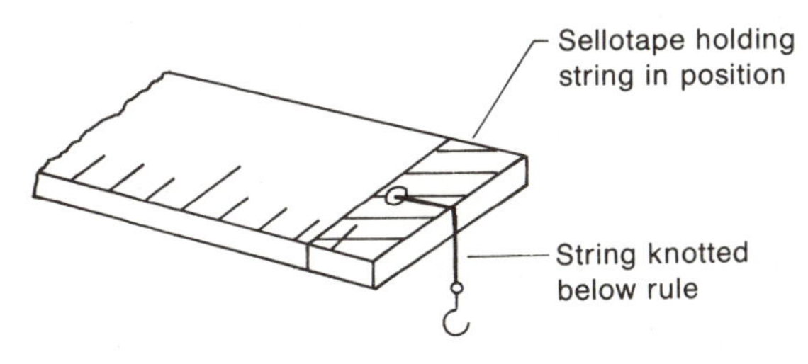

Fig. 17.

- G-clamp and 2 rectangular blocks of wood to provide a clearly defined length of cantilever
- 2 100 g and 2 50 g masses, each with a loop of thread tied to it for hanging from the hook (equivalent slotted masses can be used)
- half-metre rule
- retort stand, clamp and boss
- length of wire or pin and Plasticine or Blu-tack for fixing it to the end of the rule to act as a fiducial mark
- plane mirror to help in reading the position of the fiducial mark against the rule

(JMB 1983)

17D Oscillations of a hacksaw blade

In this experiment candidates will be asked to measure the period of the vibrating system illustrated in Fig. 18 for values of x from 0.15 m up to 0.30 m (that is, up to the highest practical value for x). Measurement of the dimensions of the vibrating system and these results are then used to determine the Young modulus E for the material concerned. In a subsidiary experiment the period for the vibrating system arranged as in Fig. 19 is to be measured.

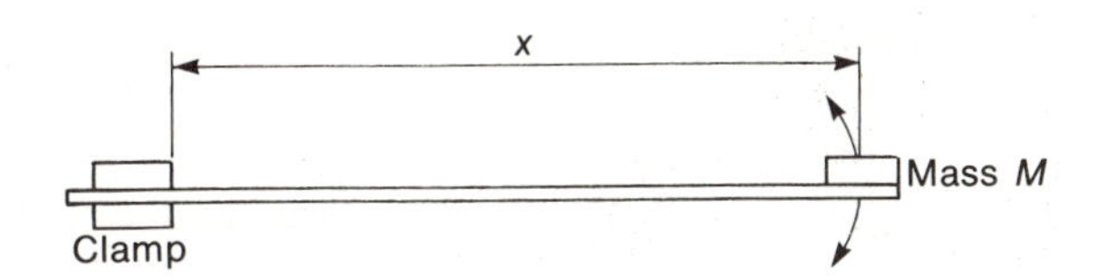

Fig. 18. Plan view.

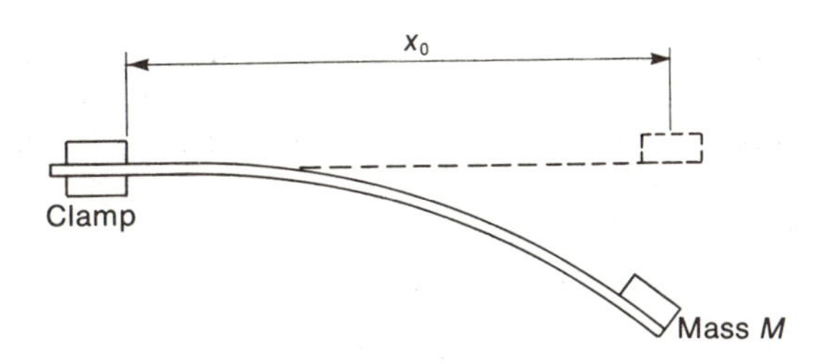

Fig. 19. Side view.

APPARATUS REQUIRED

- mass M, 100 g to 140 g, the value of which is to be given to candidates, which is to be securely fixed to the hacksaw blade so that its centre of mass is at about the same position as the hole in the hacksaw blade, and **must not be removed**
- 2 smaller masses may be used to make up to this total mass
- hacksaw blade, length 300 mm (which will typically have a cross-section of 12 mm × 0.7 mm)
- 300 mm rule
- micrometer or vernier callipers.
- stopwatch, or stopclock, or suitable digital watch
- stand and suitable clamp (or any other suitable arrangement) for clamping one end of the hacksaw blade with protruding lengths from 0.15 m to 0.30 m

The hacksaw blade will be horizontal with the face of the blade vertical in the main experiment and with the face of the blade horizontal in the subsidiary experiment. Teachers should explain to candidates how the clamping arrangement provided is intended to be used. Candidates are expected to set up the arrangement themselves.

The experiment should give the accepted value of the Young modulus E for steel.

(OLE 1983)

17E Extension of a spring

Candidates will be required to investigate the extension of a spring beyond its elastic limit.

APPARATUS REQUIRED

The following apparatus is required for all groups. The items should be provided for the candidate as listed and should *not* be further assembled by the supervisor.

- steel spring having about 25–30 turns of approximately 15 mm mean diameter (e.g. Nuffield-type Expendable Steel Spring: Griffin and George or Philip Harris. Each candidate will require a new spring. Some spare springs should also be available
- half-metre or metre rule
- 2 stands with bosses and clamps
- masses to cover the range 0–400 g in steps of not more than 20 g
- slotted weight hanger or scale pan
- card on which is written: 'Radius, *R*, of the spring = . . .'. *R* should be given to three significant figures. (To find 2*R*, measure the outside diameter of the spring and subtract the diameter of the wire.)

(L 1980)

17F Variation of the flow of water through a capillary tube with temperature

Candidates will be asked to investigate the variation of the viscosity of water with temperature by determining the time for a fixed volume to flow through a capillary tube in a siphon arrangement.

APPARATUS REQUIRED

- capillary tube, length about 100 mm and internal diameter 0.8 (± 0.2) mm
- beaker, 500 ml
- thermometer, 0–110 °C
- stirrer
- siphon tube, length about 250 mm and internal diameter about 8 mm
- screw clip
- rubber tubing
- suitable stands and clamps
- metre rule
- measuring cylinder or flask for collecting 50 cm^3 of water
- Bunsen burner, gauze and tripod
- stopwatch or stopclock

Candidates will require one litre of freshly boiled water which has been allowed to stand overnight in a refrigerator.

The supervisor should explain to candidates how the siphon is operated.

(OLE 1980)

OSCILLATIONS AND WAVES

17G Oscillations of a bifilar pendulum

Candidates are to investigate the oscillations of a bifilar pendulum.

APPARATUS REQUIRED

- 2 half-metre rules
- metre rule
- 2 retort stands and clamps
- stopwatch or stopclock
- supply of strong thread, scissors and Sellotape

The supervisor must ensure that all candidates have their pendulums swinging in the correct mode of oscillation. A sample set of data should be enclosed with the scripts.

(O & C 1982)

17H Oscillations of a pendulum against a knife edge

Candidates will be required to investigate the properties of a simple pendulum whose swing is interrupted by a fixed obstacle.

APPARATUS REQUIRED

- 2 retort stands of height sufficient to support an object 60 cm above the bench
- clamps and bosses
- knife edge (e.g. triangular file, blunt knife or bevelled edge of ruler)
- mass for pendulum bob (e.g. 100 g mass)
- strong thread (about 1 m long) to support bob
- 2 identical rectangular pieces of metal or wood between which the thread will be clamped to form a well-defined point of suspension
- metre rule
- stopwatch or stopclock
- suitable fiducial mark

(JMB 1981)

17I Oscillations of a spring system

Candidates will be required to measure the displacement of a Y-shaped system of springs for the application of various loads. They will then be asked to measure the time period of the mass-spring system for different loads.

APPARATUS REQUIRED

- 4 'expendable' springs. The type supplied by Griffin and George and Philip Harris are about 20 mm long (excluding hooks) and have a constant of about 30 N m^{-1}. Any springs that approximate to this specification are suitable. Each set of springs should be inspected at the end of the experiment and replaced if necessary

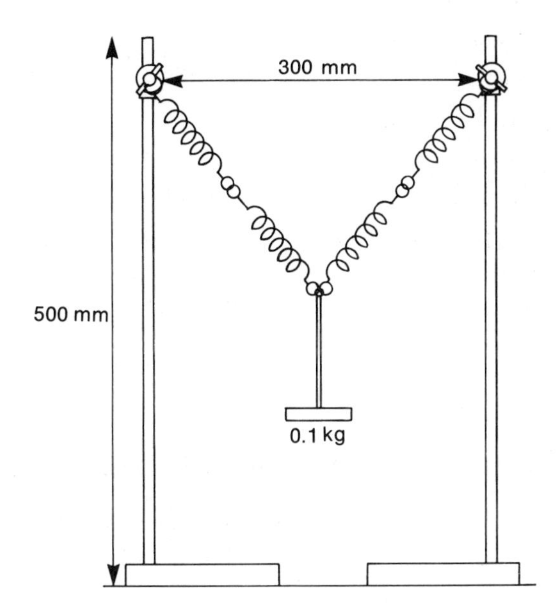

Fig. 20.

- load carrier of mass 0.1 kg
- slotted weights (or something similar) that will enable the candidate to load the spring system in steps of 0.1 kg to 0.5 kg
- 2 retort stands of height at least 0.5 m
- some means of attaching the end of the springs to the retort stands. Bosses can be fixed to the stands, and the hooks of the springs can then be attached to the bosses
- some means of fixing the positions of the stands. G-clamps or large masses are suitable
- fiducial mark
- stopwatch or stopclock
- half-metre rule (or metre rule)

The 2 stands should be fixed so that the rods are about 300 mm apart (Fig. 20). The 4 springs should be connected 'in series' each of the free ends of the combination being fixed near to the top of the stands at the same height above the bench. The load carrier of mass 0.1 kg should be attached to the centre of the spring system as shown in the diagram. The load should be such that the candidate can increase its mass in steps of 0.1 kg. It must be ensured that the stands are sufficiently tall so that, with maximum load, the load carrier does not touch the bench.

(AEB 1983)

17J Oscillations of a loaded test tube

Candidates will be asked to investigate the oscillations of a loaded test tube in water.

APPARATUS REQUIRED

(per set of apparatus unless otherwise specified).

To be supplied by the Centre

- 24 mm or 25 mm test tube, about 180 mm long
- sufficient lead shot, about 4 mm diameter, to load the test tube down to the rim when floating in water. The shot should be supplied in a shallow container
- millimetre scale, minimum length 300 mm
- glass beaker or jar, minimum diameter 100 mm, in which the test tube can be floated to its full depth.
- supply of clean water, sufficient to fill the beaker
- set square
- stopwatch or stopclock
- cloth or absorbent paper to mop up spillages
- vernier callipers. This item will have only occasional use and so one to a group of four candidates will suffice

N.B. The apparatus is to be supplied unassembled.

To be supplied by the Syndicate

Nil.

(C 1979)

17K Damped oscillations of a half-metre rule

In this question candidates will be asked to carry out observations of the damping of the oscillations of a compound pendulum, using the apparatus illustrated in Fig. 21. The results will be used to determine a value for the viscosity of water.

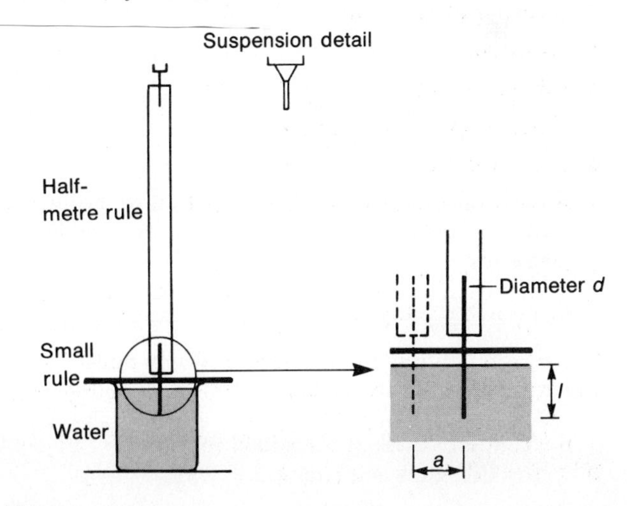

Fig. 21.

APPARATUS REQUIRED

- half-metre rule (labelled with its mass M in kilograms)
- 22 SWG (0.71 mm diameter) wire, 100 mm length and a separate length for diameter measurement
- beaker to accommodate oscillations of amplitude a of 30 mm with the wire immersed a length l of 50 mm (see Fig. 21); 80 mm internal diameter is marginally preferable to 65 mm internal diameter
- 150 mm rule, to measure depth l and amplitude a
- micrometer, to measure diameter d of wire
- stopwatch, stopclock, or suitable digital watch
- tall retort stand, boss and clamp
- bulldog clip and thread, for suspension arrangement
- water at room temperature (value of density, 1000 kg m^{-3}, is to be given to candidates)

The 100 mm length of wire is to be fixed (by the teacher) with sticky tape along the centre of either of the broad faces of the rule so that 60 mm is protruding. It is important that the oscillations of the rule should be in the plane of the broad face and this is accomplished as shown in the suspension detail of Fig. 21. The suspension arrangement and its purpose should be explained to candidates.

The significance of $\log_e$ and how its values can be found should be explained to candidates.

(OLE 1984)

17L Vibrations of a vertical wire under tension

Candidates are to investigate the resonant vibrations of a vertical wire hanging under various tensions. A diagram and description of the apparatus arrangement are given on the question paper.

APPARATUS REQUIRED

- retort stand (at least 50 cm long), G-clamp, 2 clamps and bosses
- 2 m of 0.4 mm diameter (28 SWG) copper wire
- any large horseshoe magnet, or 2 Magnadur magnets and 1 steel yoke (as used in the Nuffield O-level Westminster kit) are suitable
- low voltage a.c. supply and resistor to limit current to about 2 A through wire. This can be achieved by using a 12 V transformer with a 12 V, 24 W light bulb
- metre rule
- masses from 0 to 1000 g, e.g. 1 set slotted masses 100 g and 1 set of slotted masses 10 g and one 5 g mass. If a scale pan is required its mass must be given to the candidate (N.B. the scale pan must be so constructed that the lower end of the wire acts as a nodal point, not the centre of mass of the added masses)
- 2 crocodile clips and connecting wire

(O & C 1982)

GEOMETRICAL AND PHYSICAL OPTICS

17M Measurement of the refractive index of a liquid

Candidates are to study the refraction of light by a beaker of liquid acting as a cylindrical lens, and to find the refractive index of the liquid.

APPARATUS REQUIRED

- lampbox with line filament bulb (e.g. 12 V, 24 W) mounted so that the filament is vertical
- 250 cm^3 glass beaker of external diameter approximately 7 cm
- 6 squares of thin opaque card each of side 10 cm
- focusing screen
- metre and half-metre rules, scissors or single-edged razor blade
- water (100 cm^3) and liquid labelled X (100 cm^3)
- packing must be available to enable candidates to mount the lamp filament and lens in the same horizontal plane
- liquid X should be propane-1,2,3-triol (glycerol)

With the scripts the supervisor should enclose (a) the sets of masks used by each of the candidates, and (b) a measurement of the refractive index of the sample of liquid X used.

(O & C 1982)

17N Refraction by a cylindrical lens

In this experiment candidates will be required to trace rays through a beaker filled with water to a depth of 40 mm (see Fig. 22). In a subsidiary experiment the no parallax position for an object in front of the water-filled beaker with a mirror behind it is to be found (see Fig. 23)

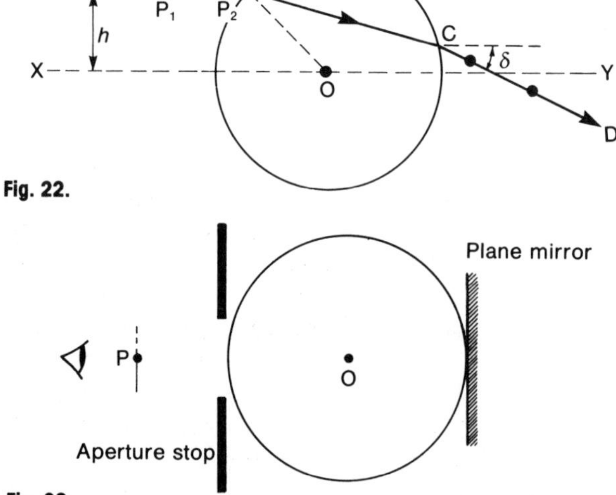

Fig. 22.

Fig. 23.

APPARATUS REQUIRED

- clean beaker, about 80 mm in diameter (mask bottom 25 mm of beaker and check uniformity of diameter from height 25 mm to 40 mm)
- squared paper, size A4
- board or cork base, for taking pins, size A4 or larger
- 4 drawing pins for pinning paper to board
- 4 optical pins, 50 mm long
- light source to illuminate the pins may be provided but is not essential
- 300 mm rule, pair of compasses, protractor, sharp pencil (candidates should be told to bring their own for the examination)
- plane mirror, to be mounted vertically adjacent to beaker
- mask with vertical slit about 10 mm wide and 40 mm high, to be mounted adjacent to beaker
- calipers to measure diameter of beaker
- supply of bubble-free water
- small block with an optical pin mounted horizontally of about 30 mm and protruding about 10 mm so that a face of the block gives the position of the pin (see Fig. 23)

(OLE 1984)

17O Use of displaced images to determine the refractive index of a block

Candidates will be asked to measure the displacement of the real image which occurs when the block of material is inserted between the lamp object and the lens, and to calculate the refractive index of the material.

APPARATUS REQUIRED

- 2 m optical bench (or equivalent)
- small filament lamp and supply
- converging lens, about 150 mm focal length, in suitable holder
- white screen
- red filter
- glass or Perspex block (with large faces clear) of thickness 15–20 mm
- metre rule
- vernier calipers

(OLE 1979)

17P Measurement of the focal length of a converging lens using conjugate images

Candidates will be required to locate the position of one of the principal foci of a converging lens by the 'plane mirror' method.

They will then be required to produce the two possible images for the particular separation of the object and the image.

APPARATUS REQUIRED

- converging lens with focal length about 15 cm
- plane mirror large enough to cover the lens
- illuminated object, labelled O. This would consist of a white screen with a circular hole about 2 cm in diameter with cross wires
- white image screen, labelled I
- holders for the above
- metre rule
- Plasticine or similar material

The metre rule should be fixed to the bench; Plasticine should suffice. The supervisor should ensure that the ambient lighting conditions are suitable for the experiment.

(AEB 1985)

17Q Measurement of the focal length of an inaccessible lens

Candidates are to locate and measure the focal length of a lens concealed in a tube.

APPARATUS REQUIRED

- 1 10 cm focal length converging lens, secured 3 cm from one end of a 10 cm long tube of any suitable material. The position of the lens must not be apparent on external examination of the tube, and both ends of the tube must be covered with thin, clear, plastic sheet (e.g. Clearwrap). The end of the tube nearer the lens to be labelled A
- light box or other source of illumination
- ground glass screen with vertical stand
- vertical screen at least 10 cm square
- means to arrange light source, glass screen, and image screen at similar height
- black stiff paper of same size as ground glass
- mm graph paper of same size as image screen
- scissors, Sellotape
- metre rule

Supervisors should send a record of the lens used by each candidate, with its focal length and position of its centre in the tube measured from end A, each accurate to 1 mm.

A sample set of data, with d and f evaluated, should also be forwarded with the scripts.

(O & C 1980)

17R Measurement of the absorption of light in glass using an LDR

In this question candidates will be asked to investigate how the light transmitted varies with the number of microscope slides (see Fig. 24). The light-dependent resistor (LDR) with a constant voltage supplied has a current directly proportional to the light intensity. The light output from the lamp is kept constant after adjustment at the start of the experiment to a suitable value.

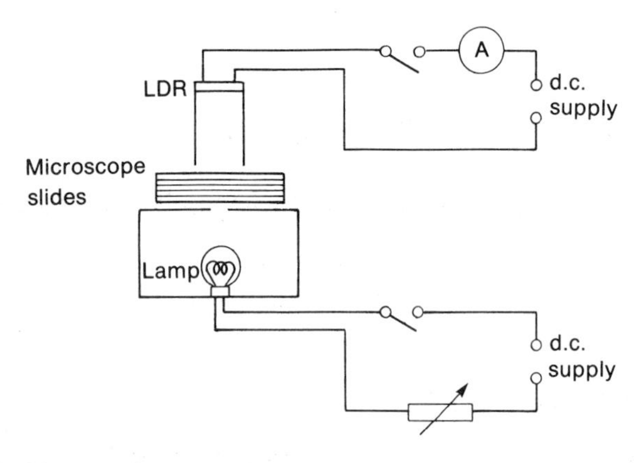

Fig. 24.

APPARATUS REQUIRED

- light-dependent resistor LDR (type ORP12, RS Components stock no. 305–620)
- ammeter, 10–25 mA f.s.d.
- voltage supply for the LDR, 6–10 V which should not change significantly (say, ±3%) as the reading of the LDR current changes from maximum to half maximum
- stand and clamp, for holding LDR tube
- 12 microscope slides
- ray box with 12 W to 24 W lamp, or equivalent, to provide well shielded source of light
- voltage supply for lamp, variable (or suitable rheostat in addition)
- 2 switches; electrical leads
- dark cloth or other means of providing a light shield

The LDR is to be mounted in a suitable light-tight cardboard tube which is held in a fixed position with respect to the lamp. The arrangement is intended to give a deflection (70% to 90% f.s.d.) of the ammeter with no microscope slides in place which reduces to about half this value with the 12 microscope slides in place. The lamp supply must be adjusted as necessary to achieve this. Note that 200 mW is the maximum power dissipation for the LDR, and that a typical dark current (that is, no illumination) of the LDR is 0.02 mA.

You will need to carry out tests to find a suitable arrangement for geometry of the lamp/LDR set-up using the apparatus in your centre. The microscope slides may be supported on a glass plate or suitable mounting made of cardboard.

The apparatus should essentially be set up ready for the candidates to use. They will need to switch on the supplies to the lamp and LDR, and are expected to adjust the lamp supply and position of the LDR as necessary so that the ammeter gives a suitable deflection (say, 90% f.s.d.) with no microscope slides in place. They will also need to check that with the lamp switched off the dark current of the LDR is negligible.

(OLE 1984)

17S Measurement of the wavelength of light using a diffraction grating

Candidates will be asked to calibrate a diffraction grating and use it to measure the range of wavelengths emitted by a white light source and the wavelengths transmitted by two filters.

APPARATUS REQUIRED

- suitable 'white' and sodium light sources and slit
- drawing board and white paper to cover it
- drawing pins
- diffraction grating or replica containing between 300 and 600 lines per mm
- 2 colour filters marked A and B. They should be complementary so that very little light is transmitted when they are superposed (e.g. magenta and primary green)
- 3 or 4 sighting pins
- protractor

(O & C 1981)

17T Measurement of the optical rotation of polarised light

In this experiment the apparatus shown in Fig. 25 is to be set up ready for candidates to make observations.

Candidates will be asked to determine the angular rotation of plane-polarised light after passage through 100 mm of solutions of different concentrations for (A) green light and (B) red light. In a subsidiary experiment the concentration of an unknown solution is to be found.

a

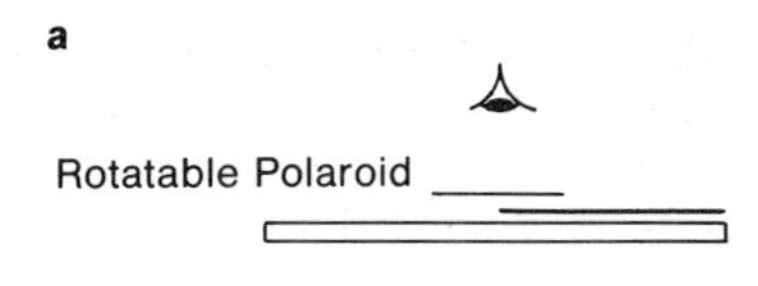

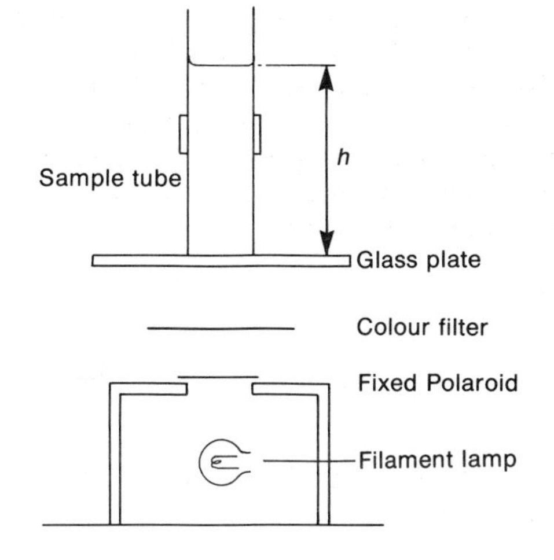

b

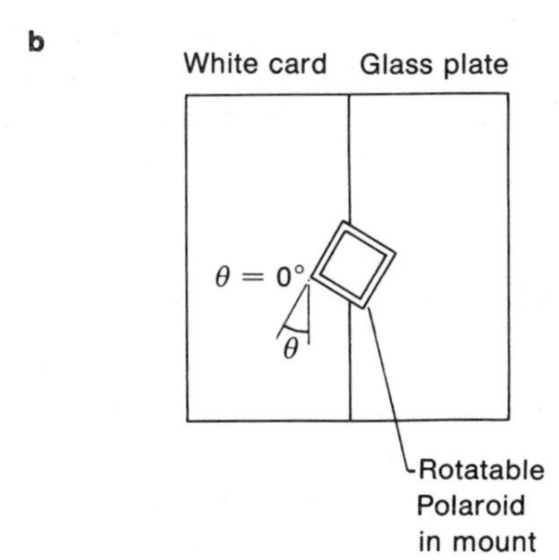

Fig. 25.

APPARATUS REQUIRED

- suitable bright source of light such as a 12 V 24 W filament lamp, with the filament on the axis of the optical system
- fixed Polaroid which **must not be moved**, about 25 mm × 25 mm
- rotatable Polaroid in a 35 mm slide mount
- red and green colour filters
- 2 pieces of glass plate, about 100 mm × 100 mm
- 4 pieces of thin white card about half the size of the glass plate
- Sellotape
- flat-bottom sample tube at least 100 mm high (that available from Philip Harris is very suitable)
- 300 mm rule
- protractor
- sharp pencil
- stand and clamps as necessary
- sugar solutions of known concentrations, labelled with letter and value of c: X, $c = 0.20$, 25 g sugar in 100 g water; Y, $c = 0.33$, 50 g sugar in 100 g water; Z, $c = 0.50$, 100 g sugar in 100 g water
- unknown solution, labelled S **only**, 70 g sugar in 100 g water

Rotations expected for 100 mm length of solutions X, Y and Z are 13°, 22° and 34° for red light and 19°, 32° and 48° for green light respectively.

(OLE 1983)

THERMAL PROPERTIES OF MATTER

17U Measurements on the cooling of borax solutions

Candidates will be asked to investigate the cooling of borax solutions in test tubes of differing dimensions.

APPARATUS REQUIRED

(per set of apparatus unless otherwise specified)

a) To be supplied by the Centre

- thermometer, 0–100 °C, graduated in degrees or half-degrees
- 2 test tubes, one nominally 18 mm diameter and the other nominally 25 mm diameter, both about 150 mm long
- cork to fit each test tube, bored with a central hole for the thermometer and with a slot to take a stirrer
- wire stirrer to fit each test tube
- retort stand and clamp
- hot solution of borax (sodium tetraborate, $Na_2B_4O_7.10H_2O$), concentration 40 g to 42 g of the *hydrated* (i.e. **not** 'calcined') salt per 100 g of water; sufficient for both tubes to be about half full
- millimetre scale, minimum length 150 mm
- watch or clock for timing to the nearest 5 s
- vernier callipers, suitable for measuring the diameters of the test tubes
- thermometer for measuring room temperature.

The vernier callipers will have only occasional use and so one to a group of four candidates will suffice. The thermometer may be a single thermometer positioned at a convenient point in the laboratory.

Under draughty conditions, candidates should be supplied with a large beaker or can suitable for surrounding their apparatus. This should be labelled 'draught shield'.

A supply of cloth or absorbent paper should be available near the water baths for drying the tubes and for mopping up spillages.

PREPARATION

The tubes, fitted with corks and stirrers and containing the solution, are to be supplied in a water bath with the solution and the bath already at a temperature of at least 95 °C. The larger tube should have the thermometer inserted and the other should have its central hole plugged with a little cotton wool.

b) To be supplied by the Syndicate
Nil.

(C 1979)

17V Measurement of the thermal conductivity of glass

Candidates will be required to measure the thermal conductivity of glass using the rate of heat flow through the wall of a test tube.

APPARATUS REQUIRED

(per set of apparatus unless otherwise stated)

a) To be supplied by the Centre

- beaker or wide-mouthed jar, capacity about 1 dm^3
- large thick-walled test tube, diameter about 25 mm
- thermometer suitable for measuring temperatures in the range 0 °C to room temperature, which can be read to better than 0.5 °C. A 25 cm, 0–100 °C thermometer will suffice
- stop clock or running wall clock with easily visible seconds hand.
- supply of ice
- vernier calipers
- retort stand and clamp
- double stirrer to fit the test tube (see *Preparation* below)
- cork to fit the test tube, bored centrally to fit the thermometer and with a side slot for the stirrer
- supply of tepid water, temperature about 30 °C

The vernier calipers may be supplied one to three or four candidates.

b) To be supplied by the Syndicate
Nil.

PREPARATION

The large beaker should be supplied to the candidate full of a water–ice mixture at 0 °C. Extra ice may be required to keep the temperature at 0 °C throughout the experiment.

The double stirrer is intended for stirring the contents of the test tube and the liquid outside the test tube. It may be constructed of wire as shown in Fig. 26. The length of the arms of the stirrer should be such as to allow the small loop to reach to the bottom of the test tube.

(C 1981)

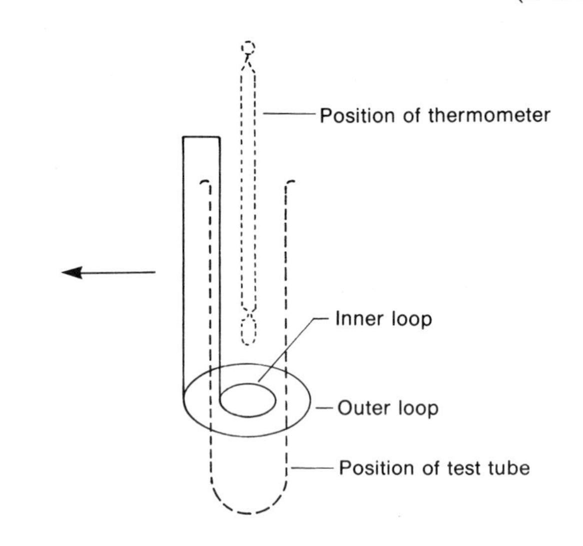

Fig. 26.

17W Measurements on a vapour in an enclosed space

Candidates will be asked to measure the atmospheric pressure from observations of the vapour of a volatile liquid in an enclosed space.

APPARATUS REQUIRED

(per set of apparatus, unless otherwise stated).

a) To be supplied by the Centre

- 180° protractor, e.g. of 10 cm diameter
- line and load to act as a plumb-bob
- thermometer and hand lens for measuring room temperature
- half-metre rule and two rubber bands suitable for fastening the capillary tube described below to the rule
- stand, plus two bosses and clamps
- thick-walled capillary tube containing a pellet of mercury about 100 mm long which seals in a small quantity of alcohol (see below) to form a space about 100 mm long containing a little volatile liquid and air, saturated with the vapour of the liquid. It may be prepared as follows: about 300 mm of capillary tube (internal diameter about 1 mm) is partly narrowed at one end by heating it and allowing it to collapse. When cool, it is dipped, wider end down, into a tube containing mercury to a depth of about 120 mm on top of which is a layer of alcohol about 2 mm to 3 mm deep. The alcohol should be colourless industrial methylated spirit (**not** mineralised). A rubber tube and

clip on the half-closed end is used to seal the tube as it is withdrawn. The tube is laid horizontally, the clip opened, and the position of the pellet of mercury is adjusted by gently tilting the tube so that the column to be enclosed is about 100 mm long. The rubber tube is removed and the narrow end sealed off. Tubes should be stored with their open ends down. They should be arranged lying horizontally on the bench for the candidates.

N.B. The value of the atmospheric pressure must **not** be communicated to the candidates, and any barometer in the laboratory should be covered.

b) To be supplied by the Syndicate

Nil

(C 1984)

If an adjustable power supply is provided, it should be preset to 3 V and candidates instructed not to adjust this value. If dry cells are used, supervisors may where necessary solder suitable lengths of connecting wire to the cells.

The 5 kΩ potentiometer is to be used as a variable resistor and suitable connecting leads should be soldered to the appropriate two terminals. It should be set to about the centre of its range.

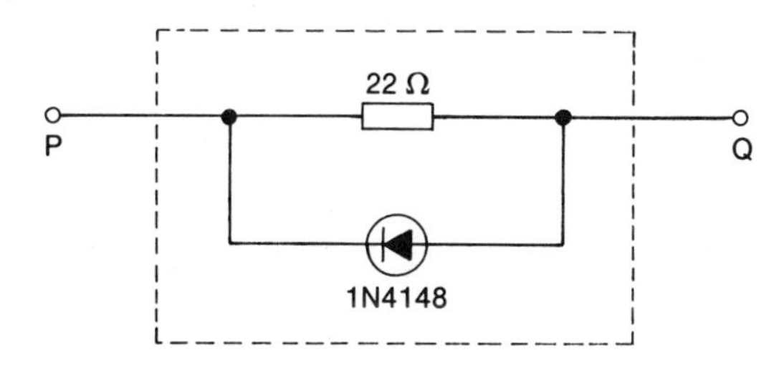

Fig. 27.

(JMB 1982)

CURRENT ELECTRICITY

17X Measurements on the electrical characteristics of a component

Candidates will be required to assemble a circuit to investigate the electrical characteristics of a component that contains a semiconductor diode.

APPARATUS REQUIRED

- 3 V fixed-value power supply (e.g. 2 dry cells or equivalent stabilised d.c. unit; supervisors should check that the dry cells, if used, are of a type (e.g. HP2) that can provide 100 mA for $1\frac{1}{2}$ hours)
- linear track 5 kΩ potentiometer (RS Components Ltd., 'Midget' type)
- pointer knob for potentiometer (RS Components Ltd.)
- switch
- 0–100 mA milliammeter
- 0–3 V or 0–5 V voltmeter
- silicon 1N4148 diode (RS Components Ltd.)
- 22 Ω 1 W high stability resistor (RS Components Ltd.)
- connecting wire
- terminals

The component to be investigated is to be made up as shown in Fig. 27 and is to be labelled clearly 'Component X'. The parts should be soldered to each other and the section within the broken line should be inside a sealed opaque container. Candidates will be instructed not to attempt to open the container. The leads P and Q should be marked clearly and should each be about 15 cm long.

17Y Measurement of the e.m.f. and internal resistance of a source

Candidates will be required to determine the e.m.f. and internal resistance of a source using a potentiometer.

APPARATUS REQUIRED

(per set of apparatus, unless otherwise specified)

a) To be supplied by the Centre

- metre slide wire
- metre rule
- 2 V d.c. steady source, e.g. an accumulator fully charged and then slightly discharged
- variable resistor suitable, when connected in series with the d.c. source and the slide wire, for reducing the potential difference across the wire to about 1.7 V, e.g. a length of resistance wire wound on a simple cardboard former and tapped off with a crocodile clip
- connecting wires, jockey and centre-zero galvanometer (with protective resistor, if appropriate)
- new U 2, 1.5 V torch battery
- A resistor of about 5 Ω, rated not less than 0.5 W, e.g. a 4.7 Ω 10% HISTAB or RS Components
- pair of crocodile clips
- length of resistance wire of resistance in the range 4 Ω to 5 Ω, e.g. 1 m of 28 SWG constantan or 24 SWG nichrome
- PVC tape, about 20 cm
- good quality voltmeter suitable for measuring the potential difference across the slide wire
- micrometer

N.B. The voltmeter and micrometer will have only occasional use so that one to about six candidates will suffice.

b) To be supplied by the Syndicate
Nil

PREPARATION

The 1.5 V torch battery, resistor and crocodile clips should be connected as shown in Fig. 28.

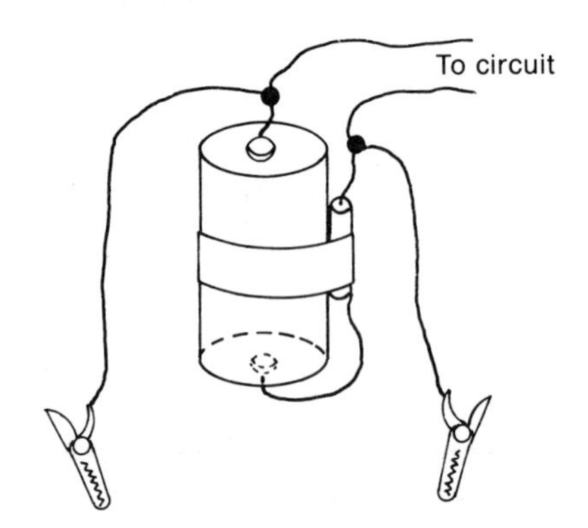

Fig. 28.

The resistor should be taped to the cell so as to hide the value of the cell e.m.f. and of the resistance. The leads to the bridge should be about 25 cm long and the leads to the crocodile clips should be about 50 cm long. The material and resistivity of the resistance wire should be communicated to the candidate in the form 'Constantan, $\rho = 4.80 \times 10^{-7}\ \Omega\ \mathrm{m}$' or as appropriate for the resistance wire supplied.

(C 1982)

17Z Measurement of the resistance of a galvanometer using a metre bridge

Candidates will be asked to carry out an experiment using a metre bridge to measure the resistance of a galvanometer.

APPARATUS REQUIRED

- metre bridge with driver cell and accessories (the total resistance of the slide wire should be less than 10 Ω)
- 2 resistance boxes covering the range 100 Ω to 1000 Ω in 100 Ω steps, *or* a selection of paired, close tolerance, 0.5 W resistors in the range 100 Ω to 1000 Ω
- galvanometer with a range of 0–1 mA or 1–0–1 mA (a standard 1 mA 100 Ω meter is suitable). If the resistance of the galvanometer is less than 100 Ω, resistance should be added in series with the galvanometer to make it approximately 100 Ω. If the value of the resistance is marked on the galvanometer, it should be obscured with tape

The driver cell should have a low internal resistance, for example a single lead-acid accumulator or a NiFe cell, or a suitable power pack.

The visiting examiner will require values for (a) the resistance of the galvanometer, (b) the e.m.f. of the driver cell.

(O & C 1980)

17AA Measurements on the characteristics of a lamp using a metre bridge

Candidates will be required to investigate how the resistance of a small lamp varies with the power dissipated in it using a metre bridge.

APPARATUS REQUIRED

(per set)

To be supplied by the Centre

- electric torch bulb (e.g. Philip Harris 2.5 V, 0.3 A*) in a miniature screw holder or with short leads soldered to it. When the bulb is in use, it is likely to be over-run for short periods but the candidate's circuit is such that the filament is very unlikely to burn out. However, some spares should be available
- battery of e.m.f. about twice the nominal voltage of the bulb: 2 fully charged lead-acid cells will be adequate for the 2.5 V bulb suggested
- standard resistor of resistance about one-third of the resistance of the bulb when brightly lit. The suggested bulb has a resistance of about 11 Ω when brightly lit and a 3.3 Ω resistor (R.S. 0.5 W Thick Film Metal Glaze) is suitable. This resistor should be labelled S. The precise resistance of this component is not critical but its resistance should be measured to the nearest 0.1 Ω and its value with an additional zero quoted on the label, e.g. the recommended resistors should be labelled 3.30 Ω
- variable resistor arrangement for varying the current through the bulb; variation is required up to ten times the resistance of the hot filament
- voltmeter with scale reading from zero to at least 1.5 times the nominal voltage of the bulb. A milliammeter in series with a resistor will do provided that the candidate is given the conversion factor. Help with interpreting readings on such an arrangement may be given without penalty

*Note. The current rating is less critical than the voltage rating.

- metre potentiometer wire with scale. The slide wire should be such that it does not become unduly hot when connected to the battery
- galvanometer with sensitivity preferably not less than 1 mm mV^{-1}
- key or switch, connecting wires and sliding contact

To be supplied by the Syndicate
Nil

(C 1981)

CAPACITANCE

17AB Discharge of a capacitor through a resistor

In this question candiates will be asked to carry out observations of the charging of an electrolytic capacitor to determine its value using the circuit in Fig. 29.

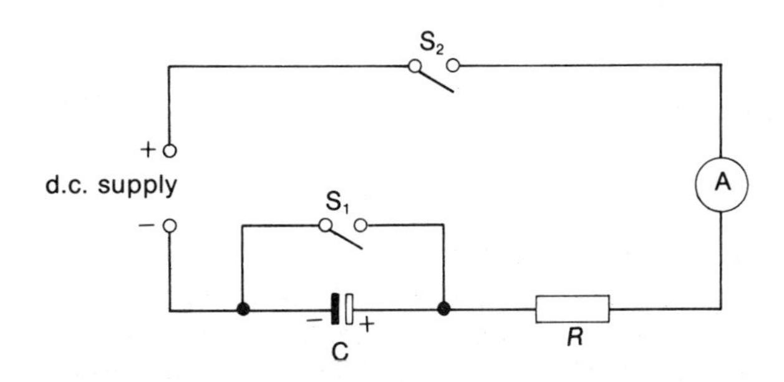

Fig. 29.

APPARATUS REQUIRED

- electrolytic capacitor, nominal value 1000 μF, 25 V working (RS Components catalogue number 103–610 is suitable), value to be masked and its polarity indicated
- resistor *R*, 100 kΩ nominal value, 0.25–1 W, value to be given to candidates
- 9 V battery or d.c. power supply
- ammeter, 100 μA f.s.d. (resistance should be small compared with value *R*)
- 2 switches
- electrical leads

Candidates are to connect up the electrical circuit in the examination.

It is intended that the circuit has a time constant $(= RC)$ of 100–150 s. The value of the resistor *R* and the voltage *V* of the supply are chosen to give a maximum current $(= V/R)$ slightly less than that for the full-scale deflection of the meter. Values of *R*, *C*, and *V* may be altered if necessary provided that the time constant remains the same and the maximum current gives the required deflection of the meter.

The invigilator is asked to check that the circuit is connected with the correct capacitor polarity.

(OLE 1984)

SEMICONDUCTOR DEVICES

17AC Measurements on the characteristics of an LED

In this experiment candidates will be asked to determine the characteristics of a red light-emitting diode (LED) and a yellow LED using the circuits shown in Fig. 30.

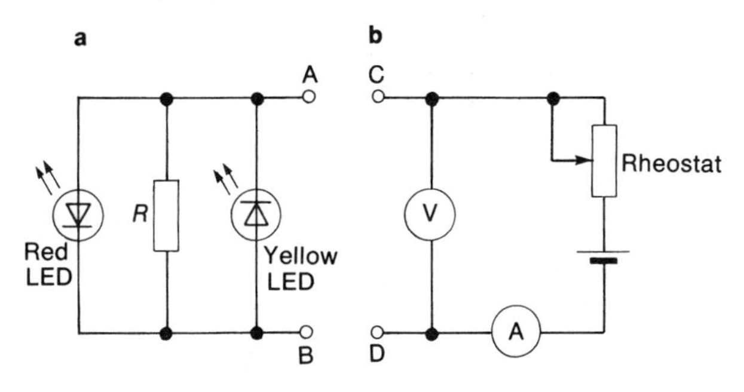

Fig. 30.

APPARATUS REQUIRED

- The circuit of Fig. 30a is to be set up before the experiment **exactly as shown** with the terminals A and B labelled. The value of *R* must be concealed. This circuit must not be altered by the candidates and this fact should be drawn to their attention.
- red LED, yellow LED, 5 mm diameter and able to take a current of 100 mA (RS Components)
- resistor *R* 47 Ω (nominal), 0.5 W, ±5%
- The circuit of Fig. 30b is to be set up by the candidates.
- suitable d.c. supply such as 9 V (PP9) battery or 6 V power pack (note that the voltage needed across the diode will not be more than 3 V and must not exceed 5 V)
- ammeter, 100 mA f.s.d.
- voltmeter, 3 V f.s.d. (or 5 V f.s.d.)
- suitable rheostat (say 2 kΩ, 1 W)

(OLE 1983)

17AD Measurements on an LED and an LDR

In this experiment candidates will be first asked to determine the characteristic of a red light-emitting diode (LED), using the circuit of Fig. 31. These results will then be used in a further experiment to investigate the change in resistance of a light-dependent resistor (LDR) with the intensity of illumination, using the circuit of Fig. 32.

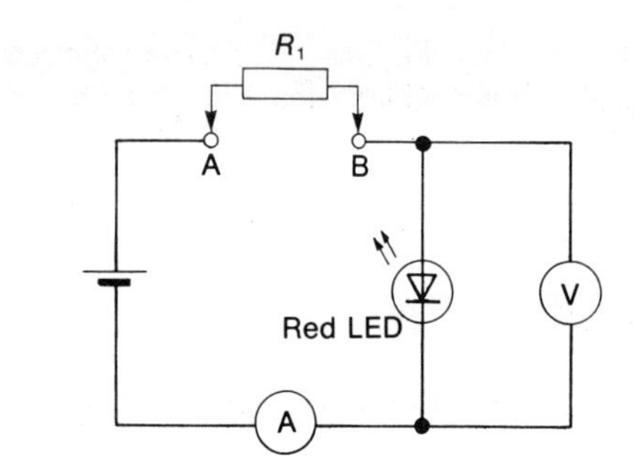

Fig. 31.

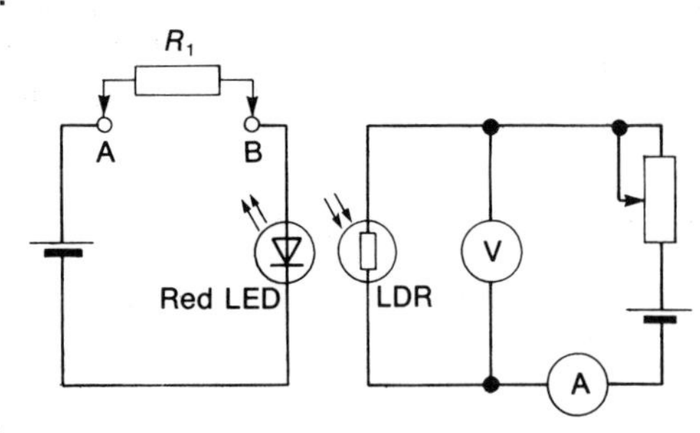

Fig. 32.

APPARATUS REQUIRED

- red LED with polarity for forward biasing clearly marked, 5 mm diameter, current up to 50 mA
- LDR, ORP12, cadmium sulphide type
- 6 resistors, 0.5 W, ±5%, labelled: R_1, 150 Ω; R_2, 200 Ω; R_3, 270 Ω; R_4, 390 Ω; R_5, 680 Ω; R_6, 1.5 kΩ
- ammeter, 50 mA f.s.d.
- voltmeter, 3 V f.s.d. (or 5 V f.s.d.)
- 2 9 V batteries (PP9 or PP3) or other suitable d.c. source (note that for the circuit of Fig. 31 the aim is to give currents of 50 mA, 35 mA, 25 mA, 18 mA, 12 mA and 6 mA by changing the resistor)
- suitable rheostat such as 2 kΩ, 1 W
- dark cloth or other means of providing a light shield

(OLE 1983)

APPENDIX 2 LEAST SQUARES FITTING OF EXPERIMENTAL DATA

1 THE PRINCIPLE OF LEAST SQUARES FITTING

The fitting procedure is based on a regression of y on x, that is the independent variable x is assumed to be error free and the errors in the dependent variable y are assumed to be random. Very few experiments satisfy this condition. Thus ideally the data should be fitted twice, once using a regression of y on x and once using a regression of x on y. This is simply achieved by interchanging the 'x' and 'y' variables on the second execution of the least squares fitting program.

For example, in the fine beam tube experiment used to determine e/m_e:

$$V_A = (eB^2/2m_e)\,r^2$$

where V_A = accelerating voltage and B = magnetic flux density producing an electron orbit of radius r.

The normal fitting would be V_A against r^2 giving a slope of $(eB^2/2m_e)$ from which e/m_e can be calculated.

For the regression of x on y you would fit r^2 against V_A giving a slope $(2m_e/eB^2)$, from which a second value e/m_e can be calculated. The difference between the two values should be less than the error in e/m_e calculated from the error in the slope.

The derivation of the equations for the slope m and the intercept on the y-axis c is based on minimising the sum S:

$$S = \sum_{i=1}^{i=N} [y_i - mx_i - c]^2 \qquad [A2.1]$$

Thus the partial differentials $\partial S/\partial m$ and $\partial S/\partial c$ with respect to m and c should both be zero for a minimum value of S:

$$\partial S/\partial m = \sum_{i=1}^{i=N} [-2y_ix_i + 2cx_i + 2mx_i^2] = 0 \qquad [A2.2]$$

and:

$$\partial S/\partial c = \sum_{i=1}^{i=N} [-2y_i + 2c + 2mx_i] = 0 \qquad [A2.3]$$

This gives two simultaneous equations in m and c:

$$\sum_{i=1}^{i=N} [mx_i^2 + cx_i - y_ix_i] = 0 \quad \text{from equation [A2.2]} \qquad [A2.4]$$

$$\sum_{i=1}^{i=N} [mx_i + c - y_i] = 0 \quad \text{from equation [A2.3]} \qquad [A2.5]$$

Using the notation:

$$S_x = \sum_{i=1}^{i=N} x_i,\; S_y = \sum_{i=1}^{i=N} y_i,\; S_{x^2} = \sum_{i=1}^{i=N} x_i^2, \text{ and}$$

$$S_{xy} = \sum_{i=1}^{i=N} x_iy_i$$

equations [A2.4] and [A2.5] become:

$$mS_{x^2} + cS_x = S_{xy} \qquad [A2.6]$$

$$mS_x + Nc = S_y \qquad [A2.7]$$

Equation [A2.7] can be rearranged to give:

$$c = (S_y/N) - m(S_x/N) = \bar{y} - m\bar{x} \qquad [A2.8]$$

where $\bar{y}$ and $\bar{x}$ are respectively the mean values of y_i and x_i. Hence the best straight line passes through the mean coordinates of all the experimental points.

Equations [A2.6] and [A2.7] can now be solved to give expressions for m and c:

$$m = \frac{NS_{xy} - S_xS_y}{NS_{x^2} - S_xS_{x^2}} \qquad [A2.9]$$

$$c = \frac{S_xS_{xy} - S_yS_{x^2}}{S_xS_x - NS_{x^2}} \qquad [A2.10]$$